From Dynamic Simulation to Optimal Design and Control of Adsorption Energy Systems

Von dynamischer Simulation zu optimalem Design und Steuerung von adsorptionsbasierten Energiesystemen

Von der Fakultät für Maschinenwesen der
Rheinisch-Westfälischen Technischen Hochschule Aachen
zur Erlangung des akademischen Grades eines
Doktors der Ingenieurwissenschaften
genehmigte Dissertation

vorgelegt von

Uwe Bau

Berichter: Univ.-Prof. Dr.-Ing. André Bardow
Prof. Robert E. Critoph, Ph. D., D.Sc.

Tag der mündlichen Prüfung: 02. Februar 2018

Aachener Beiträge zur Technischen Thermodynamik Band 12

Uwe Bau
From Dynamic Simulation to Optimal Design and Control of Adsorption Energy Systems

Von dynamischer Simulation zu optimalem Design und Steuerung von adsorptionsbasierten Energiesystemen

ISBN: 978-3-95886-216-6

Bibliografische Information der Deutschen Bibliothek
Die Deutsche Bibliothek verzeichnet diese Publikation in der Deutschen Nationalbibliografie; detaillierte bibliografische Daten sind im Internet über http://dnb.ddb.de abrufbar.

Herstellung & Vertrieb:

1. Auflage 2018

Süsterfeldstr. 83, 52072 Aachen
Tel. 0241/87 34 34
Fax 0241/87 55 77
www.Verlag-Mainz.de

ISSN: 2198-4832

Satz: nach Druckvorlage des Autors
Umschlaggestaltung: Druckerei Mainz

printed in Germany
D82 (Diss. RWTH Aachen University, 2018)

Acknowledgement

The following thesis emerged during my time as a researcher and PhD candidate at the Institute of Technical Thermodynamics at RWTH Aachen University. At this point, I would like to thank all the people who accompanied and shaped me during this time. Without all of these people, the following work would not have been possible.

First, I would like to thank my supervisor André Bardow who always supported me and gave me the inspiration to strictly follow the path of optimisation in this thesis. Additionally, his perfectionism, although sometimes frustrating, helped me to steadily improve during my time at his institute.

I would also like to thank Prof. Robert E. Critoph, Ph. D., D.Sc. for taking the time to examine this thesis and for showing great interest in the work. I also thank Univ.-Prof. Dr.-Ing. Dirk Müller for chairing the examination in a very calm and sovereign manner.

A special thanks goes to my working group, Sorption Systems Engineering: Franz Lanzerath, who was not only the group leader during this time, but who also laid the foundation of this thesis with his work and his discussions. Heike Schreiber, who made the institute a better place. Meltem Erdogan, my project buddy and favourite discussion partner concerning all philosophical questions. Stefan Graf, with whom I not only shared an office, but also plenty of fruitful discussions, at times also my frustrations, and a lot of fun. Jan Seiler, my experimental counterpart and a great friend. Andrej Gibelhaus, who not only shares my passion for simulation and optimisation, but also for Skat and Japanese music. Dominik Tillmanns and Christoph Gertig, our EST/SORP tag team. I thank all my students for their great work, namely Niklas Thielen, Lars Schumacher, Jöran Hahn, Luis Eric Olmedo, Christoph Balzer, Anna-Lena Braatz, Daniel Neitzke, Stephan Göbel, Dominik Kemetmüller, Kirstin Ganz, Mirko Engelpracht, Nils Baumgärtner, and Pooya Hoseinpoori.

I also would like to thank all the permanent staff at the institute, who keep things running smoothly in the background: Hartmut de Vries, Thomas Bungert, Djuro Dragas, Bernhard Müller, and Edgar Brauers, without whom, the experiments in this thesis would not have been possible. Gregor Migas, for supporting every strange software request. Eva Frach and Katrin Roßbruch, for keeping track of our projects and reminding me of ending contracts. Iris Wallraven, for always helping me to find my way through the organizational jungle of institute and university.

Finally, I want to thank my family and friends for always supporting me. Here, I want to name my best friend Christoph Peiniger, who accompanies me since kindergarten. My parents Hans-Georg and Renate Bau, who enabled me to follow my dreams by encouraging and assisting me in every situation. My girlfriend Anabel Brüggendieck, who gave me confidence during long weekends and short nights. Thank you!

Aachen, February 2018 *Uwe Bau*

'Give me a fruitful error any time,
full of seeds, bursting with its own corrections.
You can keep your sterile truth for yourself.'

Vilfredo Pareto, (1848 - 1923)

Contents

III Control of adsorption energy systems

Appendices

Kurzfassung

Der weltweite Heiz- und Kühlbedarf wird in den nächsten Jahrzehnten weiter steigen. Adsorptionssysteme, insbesondere Adsorptionswärmepumpen und -kältemaschinen können Teile dieses Bedarfs klimafreundlich decken, indem sie Solarwärme oder Abwärme als Antriebsquelle nutzen.

Allerdings ist die Entwicklung von Adsorptionssystemen herausfordernd, da viele Aspekte berücksichtigt werden müssen: (1) eine inherente Dynamik, (2) konkurrierende Zielgrößen, (3) eine hohe Anzahl an Designparametern, (4) ein starker Einfluss der Steuerung und (5) eine hohe Empfindlichkeit gegenüber Betriebsbedingungen. Bisher wurden diese Effekte mittels Sensitivitätsanalysen einzeln untersucht. In dieser Arbeit wird darüber hinaus auch das Zusammenspiel aus Design, Steuerung und Betriebsbedingungen untersucht und optimiert.

Die Basis für die Optimierung bilden dynamische Simulationsmodelle. Diese Modelle müssen zum einen schnell lösbar sein, zum anderen müssen sie aber auch die Effekte aller Optimierungsparameter abbilden. Zu diesem Zweck wird eine objektorientierte Modellbibliothek in der Programmiersprache Modelica entwickelt und vorgestellt.

Für die simultane Optimierung von Design und Steuerung wird ein multi-kriterielles Optimalsteuerungsproblem formuliert. Dieses Optimalsteuerungsproblem enthält die Dynamik des Adsorptionssystems durch die Einbindung eines Simulationsmodells. Zusätzlich werden Punkt- und Pfadbeschränkungen formuliert, die besipielsweise den zyklischen Betrieb des Systems sicherstellen. Das multi-kriterielle Optimierungsproblem wird mit Hilfe der ϵ-constraint Methode in viele ein-kriterielle Probleme zerlegt. Diese werden dann mit Hilfe eines effizienten Mehrschießverfahrens gelöst.

Das beschriebene Optimierungsverfahren wird benutzt, um exemplarisch das Zusammenspiel von Adsorbergeometrie, Komponentendimensionierung, Systemverschaltung und Steuerung zu untersuchen und zu optimieren. Darauf aufbauend werden weitere spezielle Fragestellungen beleuchtet, wie der Einfluss der Zielfunktion oder der Betriebsbedingungen auf die optimale Lösung. Der Trade-off zwischen Leistungsdichte und Effizienz wird mit Hilfe der erzielten Pareto Kurven diskutiert.

Die entwickelten Modelle und das Optimierungsverfahren werden genutzt, um eine modellprädiktive Regelung für Adsorptionskältemaschinen zu entwickeln und zu implementieren. Dadurch kann die Steuerung während des Betriebs optimal auf sich ändernde Randbedingungen reagieren. Darüber hinaus werden Heuristiken abgeleitet, die mit Hilfe weniger Messgrößen einen Pareto-optimalen Betrieb einer 1-Bett Adsorptionskältemaschine erlauben. Alle entwickelten Regelungen werden experimentell untersucht und bewertet.

Abstract

Worldwide heating and cooling demand will rise significantly over the next decades. Adsorption energy systems, namely adsorption chillers and heat pumps, have the potential to provide parts of this demand environmentally friendly by employing solar heat or waste heat.

Designing adsorption energy systems is challenging due to the following reasons: (1) intrinsic dynamics, (2) multi-objectiveness, (3) large variety in design parameters, (4) strong influence of control, and (5) a large impact of input parameters such as temperatures. In many studies, these effects have been investigated separately by conducting sensitivity analyses. To explore also the interactions between design, control, and input parameters, a simultaneous optimisation approach is presented and exemplified in this thesis.

Key to simultaneous optimisation are fast simulation models which capture the effects of all optimisation parameters. To quickly model new advanced adsorption energy systems, an object-oriented, dynamic-model library is developed in the programming language Modelica.

To conduct a simultaneous optimisation of design and control, a multi-objective optimal control problem is formulated. This optimal control problem includes the system dynamics which are captured in the dynamic process model. Additionally, point and path constraints are formulated, ensuring cyclic steady state. The multi-objectiveness is taken care of by employing the ϵ-constraint method. The resulting single-objective optimal control problem is solved by the efficient multiple shooting algorithm `MUSCOD II`.

The developed optimisation framework is used to rigorously analyse the interaction between adsorber-bed design, component sizing, thermodynamic cycle, and control. The framework is also used to investigate the effect of the objective function and of varying input temperatures on design and control. The multi-objective optimisation results are used to discuss the trade-off between efficiency and power density by employing the Pareto frontier.

Finally, simulation models and optimisation framework are used to set up a model predictive control (MPC). The MPC allows to identify and adjust the optimal control parameters based on varying input conditions. Based on these results, new heat-flow based heuristics are developed, which lead to a Pareto-optimal operation of one-bed adsorption chillers. Both MPC and heuristics are applied experimentally to a one-bed chiller.

List of Figures

List of Tables

Notation

Latin symbols

A	surface area	($\mathrm{m^2}$)
A	adsorption potential	($\mathrm{J\,kg^{-1}}$)
c	specific heat capacity	($\mathrm{J\,kg^{-1}\,K^{-1}}$)
c	path constraints	
C	heat capacity	($\mathrm{J\,K^{-1}}$)
d	diameter	(m)
D	Diffusion coefficient	($\mathrm{m^2\,s^{-1}}$)
h	specific enthalpy	($\mathrm{J\,kg^{-1}}$)
h	length of model stages	(s)
$\dot{H}$	enthalpy flow rate	($\mathrm{J\,s^{-1}}$)
K	permeability	($\mathrm{m^2}$)
l	length	(m)
m	mass	(kg)
$\dot{m}$	mass flow rate	($\mathrm{kg\,s^{-1}}$)
p	pressure	(Pa)
p	parameter	
q	discretised control parameters	
Q	thermal energy	(J)
$\dot{Q}$	heat flow rate	($\mathrm{J\,s^{-1}}$)
R	ideal gas constant	($\mathrm{J\,mol^{-1}\,K^{-1}}$)
s	model stage	
t	time	(s)
t	thickness	(m)
T	temperature	(K)
u	specific internal energy	($\mathrm{J\,kg^{-1}}$)
u	control variables	

v	velocity	$(\mathrm{m\,s^{-1}})$
V	volume	$(\mathrm{m^3})$
$\dot{V}$	volume flow rate	$(\mathrm{m^3\,s^{-1}})$
w	adsorbate loading	$(\mathrm{kg_{ad} kg^{-1}{}_{sor}})$
W	filled pore volume	$(\mathrm{m^3{}_{ad} kg^{-1}{}_{sor}})$
x	differential state	
X	mass fraction of water in air	$(\mathrm{kg_{water} kg^{-1}{}_{air}})$
z	algebraic sates	

Greek symbols

α	specific heat transfer coefficient	$(\mathrm{W\,m^{-2}\,K^{-1}})$
α	coefficient of thermal expansion	$(\mathrm{K^{-1}})$
β_{p}	pressure-based mass transfer coefficient	$(\mathrm{kg\,s^{-1}\,Pa^{-1}})$
β_{Darcy}	inter-particle mass transfer coefficient	$(\mathrm{kg\,s^{-1}\,Pa^{-1}})$
β_{w}	loading-based mass transfer coefficient	$(\mathrm{s^{-1}})$
β_{LDF}	intra-particle mass transfer coefficient	$(\mathrm{s^{-1}})$
δ	height	(m)
ϵ	porosity	
ϵ	epsilon constraint	
Φ	objective function	
μ	dynamic viscosity	$(\mathrm{Pa\,s})$
μ	chemical potential	$(\mathrm{J\,kg^{-1}})$
ϱ	density	$(\mathrm{kg\,m^{-3}})$

Subscripts

A	adsorber
ad	adsorbate
ads	adsorption
Ads	adsorber in adsorption mode
b	bed
bond	bonding
C	condenser
C	Carnot

c	cross sectional
Des	adsorber in desorption mode
E	evaporator
eq	equilibrium
fl	fluid
hx	heat exchanger
HR	heat recovery
IC	isosteric cooling
IH	isosteric heating
l	liquid
max	maximum
mos	model stages
min	minimum
p	particle
r	reversed
s	saturation
sor	adsorbent
tot	total
v	vapour
vap	vaporisation

Superscripts

l	liquid
v	vapour

Abbreviations and Acronyms

AHR	active heat recovery
COP	coefficient of performance
DAE	differential algebraic equation
EPT	equal phase times
HR	heat recovery
IPT	independent phase times
IVP	initial value problem

LDF	linear driving force
MHE	moving horizon estimator
MPC	model predictive control
MINLP	mixed-integer non-linear program
NLP	non-linear program
PHR	passive heat recovery
OCP	optimal control problem
QP	quadratic programming
RSD	relative state deviation
SCP	specific cooling power
SQP	sequential quadratic programming

Chapter 1

Introduction

Goal 13 of the United Nations Sustainable Goals published in 2015 reads: "Take urgent action to combat climate change and its impacts" (United Nations, 2015). The key to combat climate change is mitigating worldwide greenhouse gas emissions to possibly limit global warming to 2 °C (United Nations / Framework Convention on Climate Change, 2015). The International Energy Agency (2014) concludes "that the enhanced use of renewable energy will be crucial to decarbonise the increasing demand for heating and cooling in buildings and industry in the future" to meet the 2 °C target. In the heating and cooling sector, adsorption-based energy systems can reduce greenhouse gas emissions by using low-grade heat, such as solar or waste heat (Wang et al., 2014b).

Adsorption describes the phenomenon of fluid molecules attaching to the surface of a solid. This phenomenon can be used for thermal compression and, thus, allows to thermally drive chillers or heat pumps (Bathen, 2001). Despite the potential for efficient cooling and heating by using low-grade heat, market share of adsorption-based energy systems is small due to high investment costs and low power densities compared to conventional compression chillers and heat pumps (Li et al., 2015). The main reasons for the low power densities are the poor heat and mass transfer characteristics (Wang et al., 2014a). For the high investment costs, again the poor heat and mass transfer characteristics play a significant role since they lead to bulky systems. Additionally, manufacturing is costly due to the vacuum technology necessary for most working pairs (Wang et al., 2014a).

To improve heat and mass transfer in adsorption-based energy systems without comprising thermal efficiency, many experimental and theoretical studies have been conducted. The studies deal with the following main topics: working pairs (Aristov, 2013a), heat exchanger design (Freni et al., 2015), component sizing (Lanzerath et al., 2015), advanced thermodynamic cycles (Li et al., 2015), and control (Gräber et al.,

2011).

All of these studies investigated and improved certain aspects of adsorption chillers and heat pumps. Due to the complexity of adsorption chillers, all these aspects also influence each other. Thus, Pesaran et al. (2016) conclude in a recent review:

> "In spite of the significant research and development made in the recent years, a few certain aspects of the adsorption heat pump have not been studied comprehensively. Most of the studies, as reviewed in this work, are focused on the optimization of the adsorbent bed geometry or the operational parameters of the adsorption heat pump. However, a system-level study that includes the simultaneous optimization of the adsorber bed specifications and operational parameters of the system is lacking."

This thesis aims at closing this gap with a comprehensive analysis of an adsorption chiller by simultaneous optimisation of design and control.

Structure of this thesis

In this thesis, design and control of adsorption energy systems are improved by moving from simulation to optimisation. As representative class of adsorption energy systems, adsorption chillers are used. The developed methods and techniques can also be easily applied to adsorption heat pumps, adsorption storage systems, or desiccant systems.

In Chapter 2, the state of the art in design and control of adsorption chillers is reviewed. The chapter presents the degrees of freedom in design and control and discusses their effects on performance. Afterwards, experimental setups are reviewed which are currently used to assess adsorption chillers, followed by an overview on simulation and optimisation techniques for designing adsorption chillers.

The body of this thesis consists of three parts: (1) modelling of adsorption energy systems, (2) design of adsorption energy systems, and (3) control of adsorption energy systems.

Part I: Modelling of adsorption energy systems

In Chapter 3, an object-oriented dynamic-model library for adsorption energy systems is presented. The model library is written in the programming language Modelica. The presented library allows fast modelling of complex adsorption energy systems. The library is used to build all dynamic models used for simulation and optimisation in this thesis.

Part II: Design of adsorption energy systems

In Chapter 4, a framework is presented for simultaneous optimisation of design and control. The framework uses a dynamic model of the adsorption system and sets up a multi-objective optimal control problem. This optimal control problem is solved by an efficient multiple shooting algorithm. The results are analysed by the Pareto frontier regarding all objectives.

In Chapter 5, all steps of the framework are explained in detail and are exemplified by simultaneously optimising the adsorber-bed design and the control of a simple one-bed chiller.

The presented framework is further used to explore more design choices of adsorption chillers. In Chapter 6, the influence of component sizing is investigated. To emphasise the importance of the used objective function on the optimal design, the optimal designs for two different power density measures are determined and compared.

In Chapter 7, the influence of the working pair on design and control is explored: the optimisation conducted in Chapter 6 is repeated for a second working pair and the differences caused by different adsorption equilibria data are highlighted.

As a last example for the framework, several advanced two-bed chiller cycles are compared in Chapter 8: for each cycle, the component design and control are simultaneously optimised.

Part III: Control of adsorption energy systems

In Chapter 9, the developed modelling and optimisation techniques for adsorption chillers are used to set up a model predictive control (MPC) scheme. The MPC is used to determine optimal control parameters for varying operating conditions online. In addition, new heat-flow-based control strategies are developed. It is shown that these new control strategies lead to (near) Pareto-optimal operation of one-bed adsorption chillers. All presented control strategies are experimentally evaluated.

Chapter 10 presents a brief summary and conclusions. Furthermore, possible future research topics are outlined.

Chapter 2

State of the art in design and control of adsorption chillers

In this chapter, an overview of the state of the art in design and control of adsorption chillers is given. Adsorption chillers serve as exemplary class of adsorption energy systems. Thus, most of the described challenges in design, control, experimental assessment, simulation and optimisation also apply to other adsorption energy systems, such as adsorption heat pumps, adsorption thermal storage systems, or desiccant systems.

In Section 2.1, adsorption chillers are introduced. Adsorption is described briefly, the adsorption-chiller cycle is explained, and the main performance indicators for efficiency and power density of adsorption chillers are introduced.

In Section 2.2, the challenges in designing and operating adsorption chillers are discussed. The design choices are divided into 3 conceptual levels: the choice of the working pair, the component design, and the cycle design. As an overlaying level, the control is introduced. For each level, the general effects and potential trade-offs regarding efficiency and power density are described. Finally, the interaction between the levels is explained.

In Section 2.3, the challenges in experimentally assessing and comparing adsorption chillers are explained. The performance of experimentally tested full-scale adsorption chillers is reviewed. Based on this review, the difficulties in comparing adsorption-chiller designs due to the large influence of input conditions on performance is discussed. As an alternative to full-scale experiments, small-scale experiments are presented. Small-scale experiments are used to evaluate and characterise parts of an adsorption chiller. This information may be used by a model to estimate the performance of the full-scale adsorption chiller.

In Section 2.4, an overview on available models of adsorption chillers is given. These models range from simple thermodynamic models that estimate the maximum efficiency to sophisticated 3-dimensional models that predict the temperature distribution in an adsorbent grain. This thesis aims at optimising adsorption chillers including all levels of design. Therefore, a suitable model granularity is discussed. Finally, optimisation methods used in adsorption literature are presented and their suitability for a system-level optimisation is discussed.

In Section 2.5, the main contributions of this thesis are highlighted.

2.1 Adsorption chillers: working principle and performance indicators

Adsorption chillers are thermally-driven devices converting thermal energy into cooling power. To do so, adsorption chillers employ the physical principle of adsorption. In this section, first, adsorption is briefly explained, followed by an introduction of the simple adsorption-chiller cycle. The section ends by introducing and discussing the main performance indicators used to evaluate adsorption chillers.

2.1.1 Adsorption

Adsorption describes the process of a fluid (adsorbate) attaching to the surface of a solid (adsorbent). The reverse process, the fluid detaching from the solid, is called desorption. The combination of adsorbent and adsorbate is called working pair. In the following, the fluid is assumed to be in the vapour phase. In this section, the thermodynamic principles of adsorption are described briefly, and the nomenclature used in this thesis is introduced. The section is mainly based on Ruthven (1984).

Physical adsorption is a reversible process which is caused by van der Waals forces and electrostatic interactions. Since the energy level of the adsorbed phase is below the energy level of the fluid phase, adsorption is an exothermic process and desorption is an endothermic process.

For designing adsorption chillers, important values are the loading and the amount of heat released during adsorption or needed during desorption. The loading w is the adsorbed mass of adsorbate m_{ad} per mass of adsorbent m_{sor},

$$w = \frac{m_{\mathrm{ad}}}{m_{\mathrm{sor}}} \,. \tag{2.1}$$

The amount of heat released during adsorption corresponds to the specific enthalpy difference Δh_{ads} between vapour phase h_{v} and adsorbed phase h_{ad},

$$\Delta h_{\text{ads}} = h_{\text{v}} - h_{\text{ad}} \,. \tag{2.2}$$

The enthalpy difference Δh_{ads} is further called the specific enthalpy of adsorption.

Adsorption occurs when the vapour phase and the adsorbed phase are not in thermodynamic equilibrium, i. e. the chemical potentials of the phases are not equal:

$$\mu_{\text{ad}} \neq \mu_{\text{v}} \,, \tag{2.3}$$

and stops when thermodynamic equilibrium is reached. According to Gibbs' phase rule, the degree of freedom is 2 for pure component adsorption in equilibrium state (Frolov and Mishin, 2015), i. e. the thermodynamic equilibrium is determined by two independent variables. Traditionally, temperature T and pressure p are used as independent variables to determine the loading in equilibrium state:

$$w_{\text{eq}} = w_{\text{eq}}(p, T) \,. \tag{2.4}$$

This means, by changing temperature or pressure, the system approaches a new thermodynamic equilibrium with a different loading.

For the enthalpy of the adsorbed phase h_{ad}, temperature and loading are commonly used as independent variables (Núñez, 2001):

$$h_{\text{ad}} = h_{\text{ad}}(T, w) \,. \tag{2.5}$$

To describe the adsorption equilibrium and determine the enthalpy of the adsorbed phase, several models exist in literature. A comprehensive overview on adsorption models can be found in Duong (1998). In this work, a model proposed by Dubinin (1967) is used.

The model of Dubinin

The model of Dubinin (1967) is based on the potential theory by Polanyi (1916). Dubinin assumes that adsorption in micropores follows the mechanism of pore filling and not the mechanism of surface coverage. Dubinin (1967) assumes pore filling to be induced by an "adsorption field" in the micropore space set up by the solid body. The filled pore volume W is defined by the loading w and the density of the adsorbate $\varrho_{\text{ad}}(T)$, which is only temperature dependent:

$$W = \frac{w}{\varrho_{\text{ad}}(T)} \tag{2.6}$$

For the adsorbate density, Dubinin (1960) shows that it can be approximated by the liquid density at temperatures below the boiling point at atmospheric pressure.

As driving force for adsorption, Dubinin (1967) defines the adsorption potential:

$$A = \mu_\mathrm{l} - \mu_\mathrm{ad} = RT \ln \frac{p_\mathrm{s}(T)}{p} , \tag{2.7}$$

which can be "interpreted thermodynamically as a decrease in Gibbs free energy if the adopted standard state is the state of a bulk liquid which at a temperature T is in equilibrium with its saturated vapour at a pressure $p_\mathrm{s}(T)$" (Dubinin, 1967).

The core of Dubinin's theory is that the filled pore volume is only a function of the adsorption potential:

$$W = W(A) . \tag{2.8}$$

This function is called characteristic curve and it contains the entire equilibrium information of a working pair. The assumed temperature independence of the characteristic curve is based on experimental observations and is valid for the regarded working pairs in this thesis (Schawe, 1999; Núñez, 2001; Goldsworthy, 2014). Further discussion on this assumption can be found in Núñez (2001).

The specific enthalpy of adsorption Δh_ads can also be calculated with information contained in the characteristic curve. For this purpose, the enthalpy of adsorption is split in two parts: (1) the enthalpy of vaporisation Δh_v describing the enthalpy difference between vapour phase (v) and liquid phase (l) and (2) the bonding enthalpy Δh_bond describing the enthalpy difference between liquid phase (l) and adsorbed phase (ad):

$$\Delta h_\mathrm{ads} = h_\mathrm{v} - h_\mathrm{ad} = \Delta h_\mathrm{v} + \Delta h_\mathrm{bond} . \tag{2.9}$$

The bonding enthalpy is calculated by

$$\Delta h_\mathrm{bond} = -A + T \left.\frac{\partial A}{\partial T}\right|_w , \tag{2.10}$$

where the adsorption potential A is again the difference in Gibbs free energy and $\left.\frac{\partial A}{\partial T}\right|_w$ is the change in entropy. The partial derivative of A with respect to T can be described by:

$$T \left.\frac{\partial A}{\partial T}\right|_w = \alpha \left(\frac{\partial A}{\partial \ln W} \right)_T , \tag{2.11}$$

where α represents the coefficient of thermal expansion of the adsorbed substance and is approximated by the data for the liquid phase (Núñez, 2001).

An implementation of the model of Dubinin for the purpose of dynamic simulation and optimisation is described in Section 3.1.

2.1.2 Adsorption-chiller cycle

Adsorption is technically used for separation, drying, or thermal compression (Bathen, 2001). In this thesis, the focus lies on thermal compression which is used to run an adsorption-chiller cycle. In principle, an adsorption-chiller cycle is similar to a vapour-compression cycle: A working fluid at a low pressure level p_{E} vaporises at a low temperature level $T_{\mathrm{low}}(p_{\mathrm{E}})$. The absorbed heat necessary for vaporisation is the useful cooling power. Afterwards, the working fluid is compressed to a higher pressure level p_{C} where it condenses at a medium temperature level $T_{\mathrm{mid}}(p_{\mathrm{C}})$. To close the cycle, the pressure of the working fluid is reduced again to p_{E}. In an adsorption-chiller cycle, thermal energy at a high temperature level T_{high} replaces the mechanical energy which is used in a vapour-compression cycle to compress the working fluid.

The technical realisation of an adsorption-chiller cycle consists of three main components: an adsorber (A), an evaporator (E), and a condenser (C), which are all connected by valves (Pons and Poyelle, 1999). Adsorber, evaporator, and condenser are heat exchangers connecting the working fluid and the adsorbent to heat sources and heat sinks.

In general, the adsorption-chiller cycle operates at four independent temperature levels: the temperature level of vaporisation in the evaporator T_{E}, the temperature level of condensation in the condenser T_{C}, the temperature level of adsorption in the adsorber T_{Ads}, and the temperature level of desorption in the adsorber T_{Des}. Usually, these 4 temperature levels are reduced to 3 levels being $T_{\mathrm{low}} < T_{\mathrm{mid}} < T_{\mathrm{high}}$: The low temperature level T_{low} is the temperature of the heat source connected to the evaporator with $T_{\mathrm{E}} = T_{\mathrm{low}}$ and is set by the cooling application. The medium temperature level T_{mid} is the temperature of the heat sink and is usually set by the ambient with $T_{\mathrm{C}} = T_{\mathrm{Ads}} = T_{\mathrm{mid}}$. The high temperature level T_{high} is set by the available heat source, such as solar or waste heat. This heat source is used for desorption $T_{\mathrm{Des}} = T_{\mathrm{high}}$ and provides the necessary exergy to run the adsorption-chiller cycle.

The simple adsorption-chiller cycle consists of 4 phases: isosteric cooling, adsorption, isosteric heating, and desorption. The state changes of the working fluid occurring in each phase are explained in detail below. Additionally, Figure 2.1 shows the corresponding valve positions and heat flows. Figure 2.2 shows the thermodynamic states and heat flows of the ideal cycle in a $(\ln p)$-$(-1/T)$-diagram.

1-2 Isosteric Cooling: At the thermodynamic state 1, the minimal loading w_{min} and maximum temperature $T_{\mathrm{Des}} = T_{\mathrm{high}}$ in the cycle are reached. The valves between all components are closed, so that no adsorption or desorption takes place. The adsorbent is cooled isosterically, i. e. at constant loading, until the pressure in the adsorber $p_{\text{ad}}(T_{\text{ad}}, w_{\text{min}})$ reaches the pressure in the evaporator $p_{\text{E}}(T_{\text{low}})$. Technically, this condition may be ensured by using a flap valve between evaporator and adsorber.

2-3 Adsorption: In the adsorption phase, the valve between evaporator and adsorber is open. Adsorbate vaporises in the evaporator and flows into the adsorber, where it is adsorbed. In the ideal case, this process is isobaric since the pressure can be maintained by vaporisation. During this phase, heat $\dot{Q}_{\text{E}}$ flows into the evaporator. The adsorber needs to be cooled by the heat flow $\dot{Q}_{\text{Ads}}$ to remove the released enthalpy of adsorption Δh_{ads}. The adsorption phase stops when the minimum temperature of the adsorbent $T_{\text{Ads}} = T_{\text{mid}}$ and the maximum loading w_{max} are reached.

3-4 Isosteric Heating: Similar to the isosteric cooling phase, all valves are closed and the loading is constant. Throughout the heating phase, the pressure rises until the adsorber pressure $p_{\text{ad}}(T_{\text{ad}}, w_{\text{max}})$ reaches the condenser pressure $p_{\text{C}}(T_{\text{mid}})$. Again, in a technical realisation of an adsorption chiller, this condition may be ensured by a flap valve between adsorber and condenser.

4-1 Desorption: The valve between condenser and adsorber is open. The adsorbate streams from the adsorber to the condenser. In reversal of the adsorption phase, the adsorber is heated by $\dot{Q}_{\text{Des}}$ and the condenser is cooled by $\dot{Q}_{\text{C}}$. The desorption phase ends when the maximum temperature of the adsorbent $T_{\text{Des}} = T_{\text{high}}$ and the minimum loading w_{min} are reached.

In this thesis, the three temperature levels $T_{\text{low}} < T_{\text{mid}} < T_{\text{high}}$, further also called temperature triple, are used without loss of generality: all explanations and conclusions are also valid for $T_{\text{Ads}} \neq T_{\text{C}}$.

2.1.3 Performance indicators: efficiency vs. power density

The performance of a adsorption chillers is commonly measured by two performance indicators: efficiency and power density (Sharafian and Bahrami, 2014).

Efficiency is measured by the coefficient of performance (COP), which describes the ratio between cooling energy Q_{E} and heating energy input for isosteric heating Q_{IH} and desorption Q_{Des}:

$$COP = \frac{Q_{\text{E}}}{Q_{\text{IH}} + Q_{\text{Des}}} \,. \tag{2.12}$$

The COP only depends on the transferred amounts of energy, but not on the time which is necessary to transfer the energy.

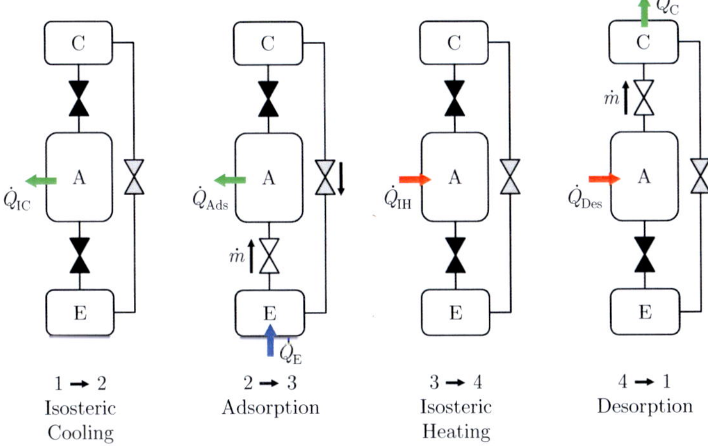

Figure 2.1: Simple adsorption-chiller cycle: Valves are connecting the 3 main components: adsorber (A), evaporator (E), and condenser (C). Valve positions (black: closed, white: open, grey return valve for liquid), mass flows and heat flows are shown for the 4 phases: isosteric cooling, adsorption, isosteric heating, and desorption (left to right). This figure is based on Lanzerath (2014).

The power density is commonly measured by a specific cooling power (SCP). A SCP relates the average cooling power $\bar{\dot{Q}}_{\mathrm{E}}$ to a reference measure for weight or volume. The average cooling power is the transferred cooling energy per cycle time $\bar{\dot{Q}}_{\mathrm{E}} = Q_{\mathrm{E}}/\Delta t_{\mathrm{cycle}}$. To maximise the average cooling power, it is important to transfer the cooling energy fast. A commonly used reference measure for weight is the adsorbent mass m_{sor} (Alam et al., 2003; Wang et al., 2005b; Qian et al., 2013; Freni et al., 2015), leading to:

$$SCP_{\mathrm{ads}} = \frac{Q_{\mathrm{E}}}{\Delta t_{\mathrm{cycle}} m_{\mathrm{sor}}} \,. \tag{2.13}$$

Other possible reference measures are the total mass of all heat exchangers, the volume of the adsorbent (Frazzica et al., 2014), the volume of the adsorber (Sapienza et al., 2011; Freni et al., 2015), or the volume of all components (Núñez et al., 2007; Lanzerath et al., 2015). The choice of the reference measure for the specific cooling power is important as it values the importance of aspects like weight or volume of the chiller design. Thus, the reference measure directly influences the preferences regarding the design. In Section 6.2 the effect of the reference measure on the preferred design is

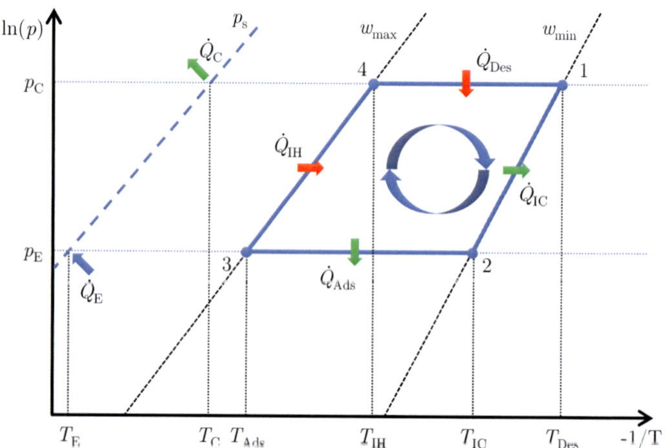

Figure 2.2: Simple adsorption-chiller cycle in a ($\ln p$)-($-1/T$)-diagram. The cycle is bound by the isotherms at minimum adsorber temperature T_{Ads} and maximum adsorber temperature T_{des}, as well as, the isobares at evaporator pressure $p_{\mathrm{E}}(T_{\mathrm{E}})$ and condenser pressure $p_{\mathrm{C}}(T_{\mathrm{C}})$. This figure is based on Lanzerath (2014).

illustrated.

Reported values for COP are in a range of 0.01 to 0.84 and for SCP_{ads} in a range of $5.7\,\mathrm{W\,kg^{-1}}$ to $820\,\mathrm{W\,kg^{-1}}$. The reported performances are further discussed in Section 2.3.1 including all references. In general, design and control have competing effects on COP and SCP_{ads} (Miyazaki and Akisawa, 2009). In the next Section 2.2, the effects of design and control on COP and SCP_{ads} are discussed. For a better understanding of the effects, the equations for COP and SCP_{ads} are rewritten.

To improve COP, the cooling energy Q_{E} has to be maximised, while minimizing the total heat input $Q_{\mathrm{IH}} + Q_{\mathrm{Des}}$. The cooling energy Q_{E} is determined by the energy balance of the evaporator: The cooling energy corresponds to the enthalpy change of the working fluid vaporising in the evaporator but is reduced by the returning working fluid from the condenser, which has to be cooled to evaporator temperature:

$$Q_{\mathrm{E}} = m_{\mathrm{sor}} \int_{w_{\min}}^{w_{\max}} \left[\Delta h_{\mathrm{v}}(T_{\mathrm{E}}) - \left(h^{\mathrm{l}}(T_{\mathrm{C}}) - h^{\mathrm{l}}(T_{\mathrm{E}})\right)\right] \mathrm{d}w \, , \tag{2.14}$$

where $\Delta h_{\mathrm{v}}(T_{\mathrm{E}})$ is the enthalpy of vaporisation at evaporator temperature T_{E} and $h^{\mathrm{l}}(T_{\mathrm{C}}) - h^{\mathrm{l}}(T_{\mathrm{E}})$ is the enthalpy difference between the liquid working fluid from the condenser and the liquid working fluid in the evaporator.

The total heat input $Q_{\mathrm{IH}} + Q_{\mathrm{Des}}$ can be divided into two parts: (1) the sensible heat needed to heat the adsorber heat exchanger fluid, the adsorber heat exchanger, the adsorbent, and the working fluid and (2) the heat input needed for desorption. The first part depends on the heat capacities C_i and the temperature difference between adsorption and desorption $T_{\mathrm{Des}} - T_{\mathrm{Ads}}$. The second part depends on the amount of desorbed working fluid $w_{\mathrm{max}} - w_{\mathrm{min}}$ and the specific enthalpy of adsorption Δh_{ads} at the desorption. The total heat input can be written as:

$$Q_{\mathrm{IH}} + Q_{\mathrm{Des}} = \int_{T_{\mathrm{Ads}}}^{T_{\mathrm{Des}}} (C_{\mathrm{hx,fl}} + C_{\mathrm{hx}} + C_{\mathrm{sor}} + C_{\mathrm{ad}}(w))\,\mathrm{d}T + m_{\mathrm{sor}} \int_{w_{\mathrm{min}}}^{w_{\mathrm{max}}} \Delta h_{\mathrm{ads}}(T_{\mathrm{ad}}, w)\,\mathrm{d}w\,. \tag{2.15}$$

Normalising Equation (2.15) by the adsorbent mass m_{sor} and replacing the heat capacity by the mass and the specific heat capacities $C_i = m_i c_i$ leads to the following expression for the COP:

$$COP = \frac{\int_{w_{\mathrm{min}}}^{w_{\mathrm{max}}} \left[\Delta h_{\mathrm{v}}(T_{\mathrm{E}}) - \left(h^{\mathrm{l}}(T_{\mathrm{C}}) - h^{\mathrm{l}}(T_{\mathrm{E}})\right)\right]\,\mathrm{d}w}{\underbrace{\int_{T_{\mathrm{ads}}}^{T_{\mathrm{des}}} (\frac{m_{\mathrm{hx,fl}}}{m_{\mathrm{sor}}} c_{\mathrm{hx,fl}} + \frac{m_{\mathrm{hx}}}{m_{\mathrm{sor}}} c_{\mathrm{hx}} + c_{\mathrm{sor}} + w c_{\mathrm{ad}})\,\mathrm{d}T}_{\text{sensibe heat input}} + \underbrace{\int_{w_{\mathrm{min}}}^{w_{\mathrm{max}}} \Delta h_{\mathrm{ads}}(T_{\mathrm{ad}}, w)\,\mathrm{d}w}_{\text{desorption enthalpy}}}\,. \tag{2.16}$$

Assuming an average loading $\overline{w}$, an average adsorption enthalpy $\overline{\Delta h}_{\mathrm{ads}}$, and introducing the overall heat capacity as $\overline{C} = \frac{m_{\mathrm{hx,fl}}}{m_{\mathrm{sor}}} c_{\mathrm{hx,fl}} + \frac{m_{\mathrm{hx}}}{m_{\mathrm{sor}}} c_{\mathrm{hx}} + c_{\mathrm{sor}} + \overline{w} c_{\mathrm{ad}}$ the COP can be written as:

$$COP = \frac{\Delta h_{\mathrm{v}}(T_{\mathrm{E}}) - \left(h^{\mathrm{l}}(T_{\mathrm{C}}) - h^{\mathrm{l}}(T_{\mathrm{E}})\right)}{\overline{\Delta h}_{\mathrm{ads}}} \frac{1}{1 + \frac{\overline{C}\,\Delta T}{\overline{\Delta h}_{\mathrm{ads}} \overline{w}}}\,. \tag{2.17}$$

The SCP_{ads} can be written as

$$SCP_{\mathrm{ads}} = \frac{\left[\Delta h_{\mathrm{v}}(T_{\mathrm{E}}) - \left(h^{\mathrm{l}}(T_{\mathrm{C}}) - h^{\mathrm{l}}(T_{\mathrm{E}})\right)\right]\,\Delta w}{t_{\mathrm{cycle,end}} - t_{\mathrm{cycle,start}}}\,. \tag{2.18}$$

To point out the importance of energy flow rates on SCP_{ads}, the cooling energy can also be expressed by the integral of adsorption rate $\dot{w}$ over time:

$$SCP_{\mathrm{ads}} = \frac{\int_{t_{\mathrm{cycle,start}}}^{t_{\mathrm{cycle,end}}} \dot{w}\left[\Delta h_{\mathrm{v}}(T_{\mathrm{E}}) - \left(h^{\mathrm{l}}(T_{\mathrm{C}}) - h^{\mathrm{l}}(T_{\mathrm{E}})\right)\right]\,\mathrm{d}t}{t_{\mathrm{cycle,end}} - t_{\mathrm{cycle,start}}}\,. \tag{2.19}$$

As Equation (2.19) shows, a high SCP_{ads} requires a high adsorption rate $\dot{w}$.

In the following section, the effects of design and control on cooling energy Q_{E}, total heat input $Q_{\mathrm{IH}} + Q_{\mathrm{Des}}$ and adsorption rate $\dot{w}$ are discussed.

2.2 Design and control of adsorption chillers

As explained in the last section, performance of adsorption chillers is commonly evaluated by two performance indicators: efficiency (COP) and power density (mostly SCP_{ads}). This leads to the necessity of a multi-objective assessment of adsorption chillers.

To systemically approach the design of an adsorption chiller, the design choices are divided into 3 conceptual levels: choice of working pair, component design, and cycle design. On top of these design levels, the control is placed as an overlaying level: For each specific adsorption-chiller design, a specific control leads to the best achievable performance. Figure 2.3 shows the conceptual levels considered in this thesis.

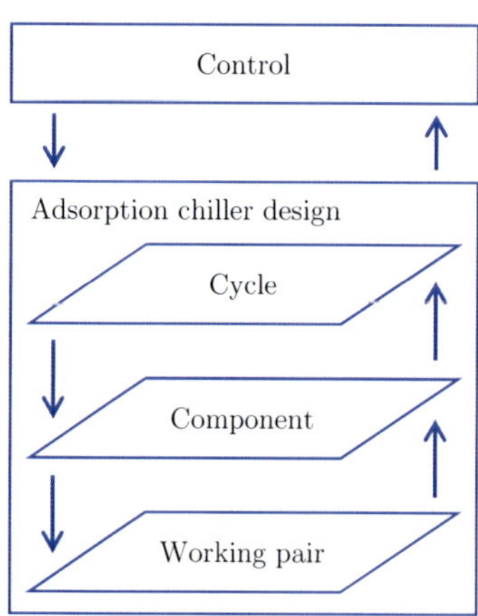

Figure 2.3: Concept of design and control levels of an adsorption chiller. The adsorption-chiller design consists of 3 conceptual levels: choice of working pair, component design, and cycle design. Control is illustrated as an overlaying level. The arrows represent interactions between the levels.

In this section, first, the design levels are presented from bottom to top: working pair, component, and cycle. Afterwards, control approaches for adsorption chillers are discussed. Finally, interactions between the levels are highlighted and discussed, pointing out the necessity of a comprehensive design approach.

2.2.1 Working pair

The used working pair (adsorbent/adsorbate) influences the performance (COP and SCP_{ads}) of an adsorption chiller by 3 inherent characteristics: its equilibrium data, its heat transfer characteristics, and its mass transfer characteristics (Aristov, 2014b; Freni, 2015). In the following, it is presented how COP and SCP_{ads} are influenced by these characteristics which leads to guidelines for the choice of an optimal working pair for a specific application.

Common adsorbents are zeolites, silica gels, selective water sorbents (SWS), metal organic frameworks (MOF), and active carbons; as adsorbate, water is most commonly used, but also methanol or ammonia are in use (Wang et al., 2009).

The COP is mainly influenced by the equilibrium data, since the equilibrium data determines the minimum and maximum loading achievable for a given set of input temperatures (T_{low}, T_{mid}, T_{high}). To improve the COP for a certain application, a high loading difference is advantageous. The higher the loading difference, the higher the ratio between cooling energy and sensible heat input, and thus, the higher the COP. Figure 2.4 shows the equilibrium data for common working pairs represented by their characteristic curves. Additionally, the ranges of adsorption potentials are highlighted for possible applications as defined in Table 2.1. By combining equilibrium data and application ranges, the maximum difference in filled pore volume ΔW_{max} can be identified, which corresponds to the maximum loading difference Δw_{max}.

Table 2.1: Temperature levels and resulting adsorption potentials of typical applications for heat pumps and chillers according to Aristov (2014b).

Application	**Temperatures**			**Adsorption potential**	
	T_{low}	T_{mid}	T_{high}	$A_{\text{min}} = RT \ln \frac{p_s(T_{\text{mid}})}{p_s(T_{\text{low}})}$	$A_{\text{max}} = RT \ln \frac{p_s(T_{\text{high}})}{p_s(T_{\text{mid}})}$
	[°C]	[°C]	[°C]	[kJ kg^{-1}]	[kJ kg^{-1}]
Air conditioning 1 (AC1)	15	30-50	60-90	130	470
Air conditioning 2 (AC2)	5	30-50	70-100	220	550
Heat pump 1 (HP1)	5	35-55	70-110	260	570
Heat pump 2 (HP2)	-15	35-55	100-140	500	800
Deep freezing (DF)	-18	30-50	100-140	500	840

As Equation (2.16) indicates, a high enthalpy of vaporisation Δh_{v} of the adsorbate is desirable to increase the COP. Finally, the specific bonding enthalpy Δh_{bond} between adsorbate and adsorbent should be low to promote the COP. Since the bonding

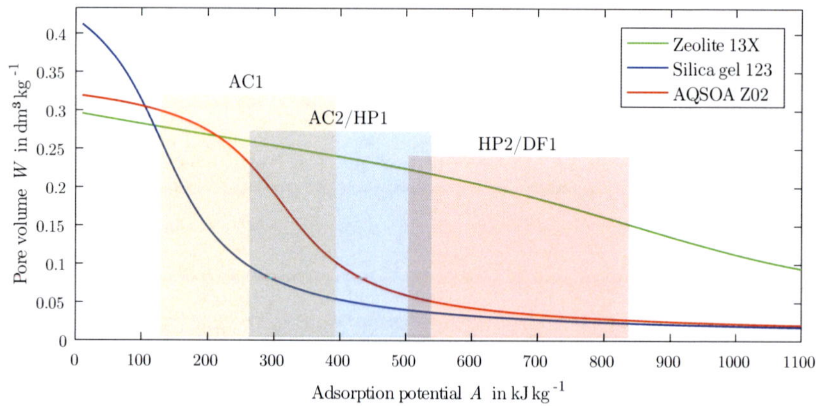

Figure 2.4: Characteristic curves for water adsorption on different adsorbents: Zeolite 13X (Núñez, 2001), silica gel 123 (Schawe, 1999), and AQSOA Z02 (Goldsworthy, 2014). ΔA-windows for different applications: air conditioning 1 (AC1), air conditioning 2 / heat pump 1 (AC2/HP1), and heat pump 2 / deep freezing 1 (HP2/DF1) (cf. Table 2.1). This figure is based on Aristov (2014b).

enthalpy is correlated to the adsorption potential as stated in Equation (2.10), it is not an independent optimisation variable.

The SCP_{ads} is influenced by equilibrium data, heat transfer characteristics, and mass transfer characteristics. The effect of equilibrium data on the SCP_{ads} is not as straightforward as on the COP. Still, Figure 2.4 allows to identify advantageous working pairs: The minimum and maximum adsorption potentials of an application set the overall driving force for adsorption and desorption respectively. The difference between the adsorption potentials set by the application and the characteristic curve of the working pair determines the specific driving forces for adsorption and desorption which directly affects the dynamics of the process. Okunev and Aristov (2014) conclude that for an optimal working pair, the characteristic curve should be a step function, for which the entire loading difference occurs within the ΔA-window set by the application. The position of the step within the application's ΔA-window defines if adsorption or desorption dynamics are faster: a step function at the left edge of the ΔA-window promotes a fast desorption, a step function at the right edge promotes a fast adsorption. How the shape of the characteristic curve influences performance is further investigated in Chapter 7.

Heat and mass transfer limitations reduce the adsorption rate and, therefore, also reduce the SCP_{ads} (Deng et al., 2011). Since these limitations are strongly connected to the adsorbent geometry in combination with the adsorber heat exchanger design, they are discussed in the following Section 2.2.2.

The discussion about the optimal working pair shows that there is no optimal working pair, but an optimal combination of working pair and application. Current research aims at tailoring known materials and synthesising new materials with a high loading difference and a step function within the application's ΔA-window. A comprehensive overview of current working pairs and ongoing developments can be found in Wang et al. (2009) and Aristov (2013a, 2014b).

2.2.2 Components

The second conceptual design level is the component level (cf. Figure 2.3). In adsorption chillers, the components are heat exchangers which connect a heat exchanger fluid to the adsorbent and adsorbate. In the following, the main challenges in designing adsorber, evaporator, and condenser are explained.

Adsorber

An adsorber is a heat exchanger which on the one hand thermally connects a heat exchanger fluid to the adsorbent and on the other hand connects the adsorbent to the working fluid in evaporator or condenser. Figure 2.5 shows the structure of an adsorber, including the adsorbent material, the heat exchanger, and the heat exchanger fluid. Also, the dominating heat and mass transfer resistances are shown.

To increase the COP, the sensible heat input has to be minimised (cf. Equation (2.16)). This can be achieved by two measures: (1) changing the heat exchanger design to reduce the ratio between heat exchanger mass and adsorbent mass, (2) changing the heat exchanger material to reduce the specific heat capacity of the heat exchanger.

To improve SCP_{ads}, a high adsorption rate $\dot{w}$ is needed. The adsorption rate is determined by the driving force (see Section 2.2.1) and all heat and mass transfer resistances. As Figure 2.5 shows, there are in principle 6 resistances in the adsorber which are connected in series (Li et al., 2015): (1) inter-particle mass transfer resistance, (2) intra-particle mass transfer resistance, (3) heat conduction in the adsorbent bed, (4) heat transfer resistance from the adsorbent bed to the heat exchanger, (5) conduction in the heat exchanger and (6) heat transfer resistance from the heat exchanger to the heat exchanger fluid. In a good design, all resistances are low and well balanced with no bottleneck present.

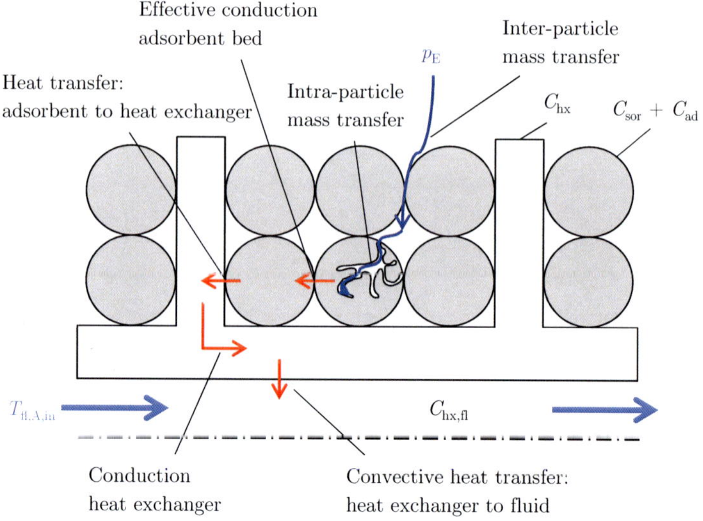

Figure 2.5: Structure of an adsorber including adsorbent material, heat exchanger and heat exchanger fluid. Main heat capacities C and heat and mass transfer resistances are labelled. Also, evaporator pressure p_E and heat exchanger inlet temperature $T_{A,fl,in}$ are shown, since they are inducing the driving potential for adsorption. The arrows relate to the case of adsorption.

The heat exchanger designs proposed in literature can be categorised in two groups (Wang et al., 2014a): (1) adsorbers using granular adsorbent pellets and (2) coated adsorbers.

Adsorbers using granular adsorbent pellets suffer most from poor heat transfer from the adsorbent bed to the heat exchanger and poor conduction in the adsorbent bed (Li et al., 2015). To overcome this problem, designs with large heat transfer areas and thin beds are beneficial. Since a large heat transfer area comes along with a high mass ratio between heat exchanger and adsorbent m_{hx}/m_{sor}, a trade-off between COP and SCP_{ads} occurs (cf. Equation (2.16)). Sharafian and Bahrami (2014) categorise heat exchanger designs and compare experimental results with respect to mass ratio, SCP_{ads} and COP. They conclude that finned tube adsorber designs provide a good compromise, but that geometrical parameters, such as fin spacing and fin height, have to be optimised.

Aristov et al. (2012a) investigate the effect of grain size and number of layers on heat and mass transfer. They conclude that for big grain sizes intra-particle mass transfer limits the system dynamics, which they call the "adsorbent sensitive" regime. For smaller grain sizes, heat transfer between adsorbent and heat exchanger limits system dynamic, which they call "grain size insensitive" regime. They substantiate their findings by investigating the effect of heat transfer surface to adsorbent mass ratio A_{hx}/m_{sor}. They find that within the "grain size insensitive" regime, a constant surface/mass ratio leads to the same dynamics independent of the grain size.

Coated adsorber designs have been proposed to enhance the contact heat transfer between adsorbent and heat exchanger. In general, coated adsorbers have enhanced heat transfer but mass transfer becomes limiting (Aristov, 2013a). To reduce the mass transfer resistance and to obtain stable coatings, a thin coating is preferred (Chang et al., 2005). Compared to granular-adsorbent-pellet designs, a thin coating leads to a higher mass ratio between heat exchanger and adsorbent, resulting in lower values for COP (Freni et al., 2015). Overviews of investigated coated adsorber designs can be found in Meunier (2013) and Li et al. (2015).

The short overview shows that designing an adsorber is a complex task due to the manifold degrees of freedom and the competing objective functions.

Evaporator and condenser

Evaporator and condenser determine the vapour pressure during adsorption and desorption, respectively. Figure 2.6 shows the structure of evaporator and condenser. The vapour pressures are determined by the heat exchanger input temperatures T_E and T_C and the heat transfer resistances. The higher the resistances, the more adsorption potential is lost in evaporator or condenser. To reduce these losses, one can either increase the heat transfer area or reduce the specific heat transfer resistance. In general, an increase of heat exchanger area is undesired because it causes an increase in weight and volume.

As Lanzerath (2014) showed for the condenser, high volume-specific heat transfer can be achieved with a helix heat exchanger design leading to a compact condenser design. For an experimentally investigated adsorption chiller, this led to a condenser accounting for 5 % of the overall chiller volume (Lanzerath et al., 2015).

For the evaporator, a compact design is more difficult. For common evaporator temperatures of 5–10°C, the corresponding vapour pressures are very low, causing the favourable nucleate boiling to require wall superheat of up to 20 K (Giraud et al., 2015). High wall superheat is undesired because it reduces the evaporator pressure p_E for a given input temperature T_E lowering the driving force for adsorption. To avoid

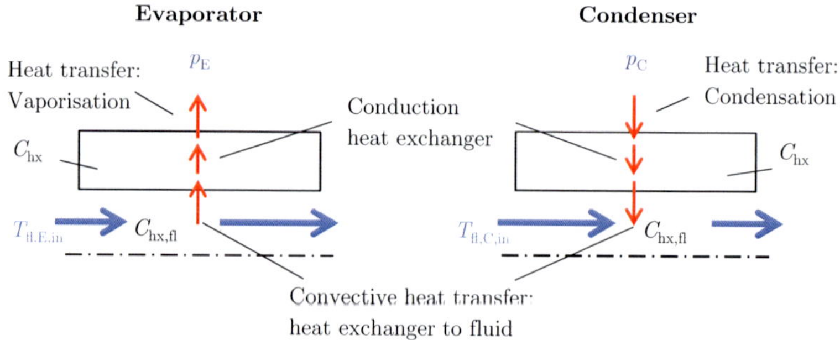

Figure 2.6: Structure of evaporator and condenser. The driving forces for vaporisation and condensation are the vapour pressure p and the heat exchanger fluid input temperatures T.

the losses due to the high wall superheat, either macroscopic structures, such as fins (Xia et al., 2009), or microscopic structures, such as coatings, are used (Lanzerath et al., 2016).

The heat transfer resistances of evaporator and condenser and the resistances in the adsorber are connected in series. In an optimal design, these resistances need to be balanced to avoid oversized components on the one hand and bottlenecks on the other hand. Therefore, a simultaneous optimisation of all components is necessary to achieve a balanced adsorption-chiller design. Since heat and mass transfer resistances do not have the same unit, it is difficult to directly compare their influence on performance. To overcome this problem, in Section 5.3.2, it is proposed to compare heat and mass transfer resistances by their loss in adsorption potential A. In Chapter 6, a simultaneous optimisation is conducted. The optimisation results are illustrated and interpreted by comparing the losses in adsorption potential caused by the resistances before and after the optimisation.

2.2.3 Cycle

The third conceptual design level is the cycle design (cf. Figure 2.3), which includes the component configuration and operating strategy. In Section 2.1.2, the simple one-bed adsorption-chiller cycle is presented using a configuration with one adsorber, one evaporator, and one condenser. In literature, many other configurations are proposed. Figure 2.7 shows two examples: an integrated modular configuration with adsorber and evaporator/condenser in one vacuum chamber (Núñez et al., 2007) and a two-bed configuration with one evaporator and one condenser (Pons and Poyelle, 1999).

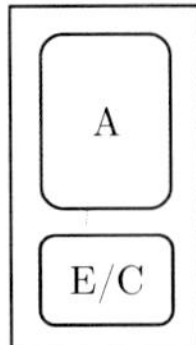

(a) Integrated one-bed configuration

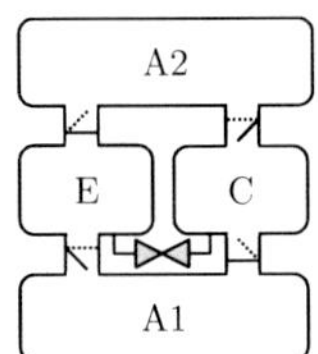

(b) Two-bed configuration

Figure 2.7: Examples of adsorption-chiller configurations: (a) Integrated one-bed configuration with single evaporator/condenser heat exchanger. Adsorber and evaporator/condenser are placed in one vacuum chamber. (b) Two-bed configuration with 2 adsorbers, an evaporator and a condenser with separate vacuum chambers each. Flap valves are used to connect the adsorbers to evaporator and condenser.

The main advantage of the integrated one-bed configuration is its simplicity: due to the fact that the heat exchangers are in one vacuum chamber, no additional vacuum valves are necessary. Also, this design can be used as a basis for a modular multi-bed design (Núñez et al., 2007).

The two-bed configuration is the most simple multi-bed configuration and has two major advantages compared to a one-bed system: (1) more continuous cooling production since one adsorber can always be in adsorption mode and (2) potentially increased efficiency and power density due to heat or mass recovery between adsorber beds. These advantages come at the cost of a higher complexity.

A number of cycle designs can be realised given a specific one or multi-bed configuration. Wang et al. (2014b) specify the cycle designs by three levels: (1) the continuity of cooling production achieved by one or more beds, (2) for multi-bed cycles, the concept to improve efficiency, e. g. heat recovery, mass recovery, or thermal wave and resulting from (1) and (2) the specific type of cycle design. Figure 2.8 gives an overview of the possible cycle designs and shows the manifold options. In this thesis, the simple one-bed cycle, two-bed cycle, and two heat recovery cycles are used to exemplify the proposed comprehensive design approach for adsorption chillers (Chapter 8).

Heat recovery schemes aim on increasing the efficiency of an adsorption chiller. They do so by transferring energy from a hot adsorber at T_{high} to a cold adsorber at T_{mid} when switching between adsorption and desorption mode. In the following, the two considered heat recovery schemes are presented: active and passive heat recovery.

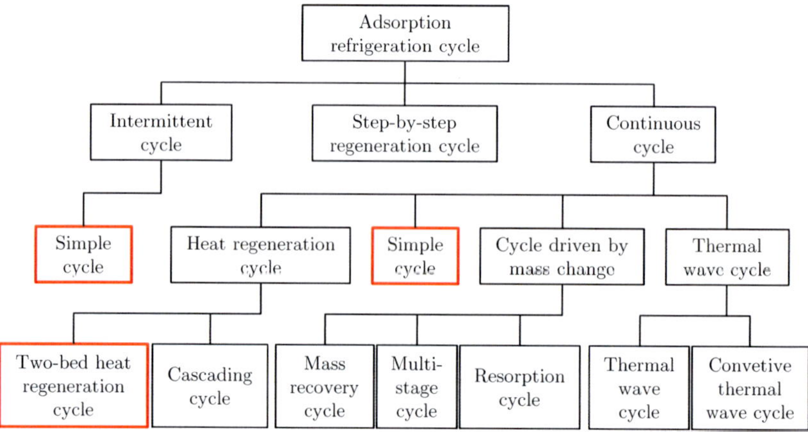

Figure 2.8: Classification of cycle designs proposed by Wang et al. (2014b). The red-framed cycles are investigated in this thesis: one-bed intermittent simple cycle, two-bed continuous simple cycle and two-bed heat regeneration cycle.

Active heat recovery

Active heat recovery schemes aim at recovering the internal energy of all adsorber parts represented by heat capacities $C_{\mathrm{hx,fl}} + C_{\mathrm{hx}} + C_{\mathrm{sor}} + C_{\mathrm{ad}}$ and also some parts of the enthalpy of adsorption (Choudhury et al., 2013). Two adsorbers are connected via a heat exchanger fluid circuit. To transfer heat from the hot adsorber (which was in desorption mode) to the cold adsorber (which was in adsorption mode). Figure 2.9 shows the active heat recovery of a two-bed system in a $(\ln p)$-$(-1/T)$-diagram. Heat recovery occurs for state transitions $1 \rightarrow 3$ and $4 \rightarrow 6$. Heat recovery is limited to the point where both adsorbers have the same temperature. Therefore, a two-bed system can only partly recover the internal energy. By adding additional adsorber beds, the recovery rate may be increased.

Active heat recovery has competing effects on efficiency and power density: By recovering internal energy, the necessary external heat input is reduced, leading to an increased COP. At the same time, heat recovery leads to slower system dynamics, because the temperature difference between two adsorbers is lower compared to using the external heat source and heat sink at T_{Ads} and T_{Des}. Thus, heating and cooling with active heat recovery is slower and reduces the SCP_{ads}.

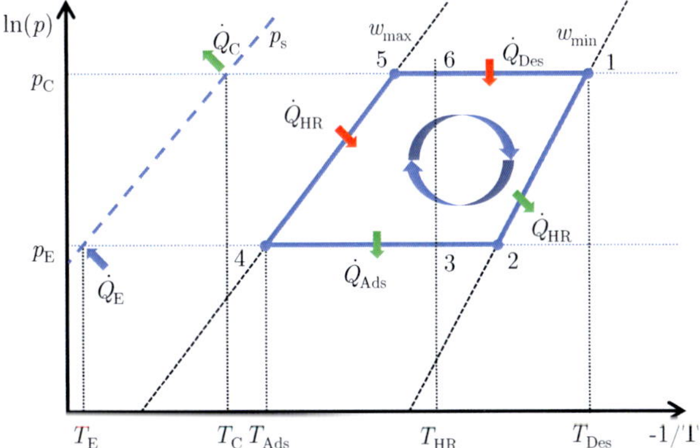

Figure 2.9: Active heat recovery scheme shown in a $(\ln p)$-$(-1/T)$-diagram. The maximum recoverable heat flow rate $\dot{Q}_{HR}$ is achieved when both adsorbers reach the equal temperature T_{HR}.

Passive heat recovery

Wang et al. (2005b) propose a passive heat recovery for a four-bed chiller. The presented passive heat recovery scheme aims at recovering the internal energy of the heat exchanger fluid in the adsorber represented by the heat capacity $C_{hx,fl}$ (cf. Figure 2.5). Passive heat recovery lowers the necessary external heat input leading to a higher COP (cf. Equation (2.16)). To achieve passive heat recovery, the valves in the heat exchanger fluid circuit are switched time-displaced. When switching from adsorption to desorption, the valve at the adsorber input is switched first, connecting the adsorber to the heat source. The output of the adsorber remains connected to the heat sink, avoiding the cold fluid to get from the adsorber to the heat source. When the cold fluid in the adsorber is displaced by hot fluid, the valve at the adsorber outlet is also switched, connecting the adsorber entirely to the heat source. The optimal time-lag between the valves can be calculated by the flow speed of the heat exchanger fluid and the volume of the adsorber heat exchanger tube.

By applying the passive heat recovery scheme, Wang et al. (2005b) could increase the COP by 30 %, while achieving the same SCP_{ads}-values.

Active and passive heat recovery schemes compete: By using the active heat recovery scheme, the temperature of the heat exchanger fluid of the adsorbers equalises, making a passive heat recovery scheme less attractive. The superior scheme is determined by the heat capacity ratio between adsorber and heat exchanger fluid and by the preferences according to COP and SCP_{ads} (Wang and Chua, 2007).

2.2.4 Control

In this thesis, control is regarded as overlaying level, since it is important in both the design phase and in the operating phase of an adsorption chiller: In the design phase, determining the optimal control is necessary to compare designs at their best operating points. In the operating phase, determining the optimal control is important to optimally run a specific adsorption chiller for the given operating conditions.

Control in the design phase

As described in Section 2.1, adsorption chillers operate cyclically, meaning they are switched between adsorption and desorption phase. The simple one-bed cycle includes the four phases: isosteric cooling, adsorption, isosteric heating, and desorption. Controlling an adsorption chiller means determining the times of each phase. For the simple cycle, the performance of a control strategy using equal adsorption and desorption phase times has been studied widely. It has been shown that there is a trade-off between COP and SCP_{ads}: Long cycles promote COP, shorter cycles yield a higher SCP_{ads} (Chua et al., 1999, 2004). Additionally, a reallocation of adsorption and desorption phase times can improve COP and SCP_{ads}. Depending on the equilibrium data and the heat and mass transfer resistances, desorption or adsorption is faster. Thus, according to the specific system's dynamics, the ratio between the phase times should be adjusted (Aristov et al., 2012b; El-Sharkawy et al., 2013; Sapienza et al., 2012).

The described effects of control on efficiency and power density have major implications when designing adsorption chillers. To ensure that only the intrinsic characteristics of designs are compared, it is necessary to compare designs only under optimal control. In this thesis, the model-based determination of optimal control is proposed within the presented comprehensive design approach.

Control in the operating phase

In the operating phase, control has to ensure that the adsorption chiller operates at its best performance for the given operating conditions. Fluctuating operating conditions, as common in solar or mobile applications, lead to continuously changing optimal phase times (Wang et al., 2008). Therefore, a control is needed which reacts on changing operating conditions quickly and which determines the optimal cycle time accurately.

The easiest way of operating an adsorption-based energy system is by setting a fixed cycle time, which stays constant during operation. This fixed cycle time may only be optimal for one specific operating point. Thus, sub-optimal operation for fluctuating input conditions is unavoidable.

Another control approach, called heuristics in this thesis, uses control criteria based on measured state variables. Schicktanz (2011) proposes the mean chilled water outlet temperature as control criterion and compared it to a control based on fixed cycle times (Schicktanz, 2011). The temperature-based control can improve SCP_{ads}, but shows to be not robust against fluctuations of the chilled water inlet temperature. Gräber (2013) proposes to compare the current cooling power with the average cooling power of the cycle. Here, the average cooling power of the cycle is the integral of the cooling power measured from the start of the cycle divided by the actual cycle time. When the current cooling power drops below the average cooling power, phases are switched. This control strategy leads to SCP_{ads}-optimal operation for a simple two-bed cycle with equal phase times for adsorption and desorption. Further details on this control strategy are provided in Section 9.2. Gräber (2013) shows that the heat-flow-based control strategy is more robust against disturbances of inlet temperatures than the temperature-based control strategy. Heuristics are easy to implement, but they do not allow for different optimality criteria and also do not guarantee optimality.

A third possible strategy to control adsorption chillers is model predictive control (MPC). MPC, frequently also called receding horizon control (Rawlings and Muske, 1993), is an online optimal-control-based method for computing stabilising feedback laws. MPC computes the feedback law by successively solving finite-horizon optimal control (open-loop) problems using the systems' state at the current time as initial value. After obtaining the optimal control for the finite horizon, the control parameters of the system are updated using the results of the obtained optimal control. At this point, a new finite-horizon optimal control loop starts with updated systems' states and updated boundary conditions. By successively solving finite horizon optimal control problems and updating control parameters, a closed-loop control strategy is obtained which steers the system towards a control goal. MPC is an established control methodology. Krstic and Smyshlyaev (2008) provide general discussion and references on MPC. In the context of adsorption chillers, MPC has 3 main advantages: (1) the control can react on changing input conditions, (2) the objective function can be chosen freely, allowing to seek e. g. for optimal SCP_{ads} or COP, (3) additional restrictions can be implemented, e. g. that the cooling output temperature should not exceed a certain value. The main disadvantage of MPC is its complexity, since an accurate model and a fast and stable optimisation procedure are needed.

In Chapter 9 of this thesis, an MPC for adsorption chillers is presented and tested experimentally for the first time. Additionally, new heat-flow-based control strategies are derived, allowing (near) Pareto-optimal operation of one-bed adsorption chillers with unequal adsorption and desorption phase times.

2.2.5 Interaction between working pair, component design, cycle design, and control

In Sections 2.2.1-2.2.4, the challenges for the levels working pair, component design, cycle design, and control are described: Additionally to the challenges on each level, these levels interact, leading to an even more complex task of designing adsorption chillers. In the following, three examples are presented, where drawbacks in one level can be compensated in another level.

Firstly, the interaction between component and cycle design is exemplified. When choosing an adsorber with a high heat transfer area, fast adsorption dynamics can be achieved. Such a design will probably lead to a high SCP_{ads}, but also a low COP since a high heat transfer area often comes at the cost of a high ratio of heat exchanger mass to adsorbent mass. To increase the COP on system level, it is possible to use a heat recovery scheme, compensating some drawbacks of the high ratio of heat exchanger mass to adsorbent mass. The heat recovery scheme would reduce SCP_{ads}, due to the time needed for heat recovery (Ng et al., 2006). Alternatively, it would be possible to choose an adsorber design with slower dynamics and a lower ratio of heat exchanger mass to adsorbent mass. For this adsorber design, the benefits of heat recovery on system level would be less. It is difficult to decide which combination of adsorber design and cycle design is favourable, since the decision requires a comparison taking into account both levels, component and cycle.

Secondly, the interaction between working pair and component design is discussed. The driving force for adsorption is set by the equilibrium data of the working pair in combination with the temperature triple of the application (cf. Section 2.2.1). The higher the driving force in adsorption and desorption, the larger the heat transfer surface needs to be in the adsorber to not limit the process. This heat exchanger surface normally comes at the cost of a higher amount of heat exchanger mass, reducing the COP and increasing the chiller's weight (Bau et al., 2016a). Therefore, it is desirable to harmonise working pair and heat exchanger design to not oversize the adsorber heat exchanger surface. This example again needs a simultaneous consideration of two design levels, working pair and component.

Thirdly, the interaction between design and control is important. As explained in Section 2.2.4, the control has a large impact on COP and SCP_{ads}. The control, where COP or SCP_{ads} achieve their maximum values depends on equilibrium data of the working and all heat and mass transfer resistances in the chiller (Aristov et al., 2012b; El-Sharkawy et al., 2013; Sapienza et al., 2012). Therefore, it is necessary to re-evaluate the optimal control for each specific design which means that control and all design levels also have to be considered simultaneously.

These three examples show that designing an adsorption chiller is a complex task. To solve this complex task, it is translated into an optimisation problem in this thesis. When design levels are optimised on their own, only a sub-optimal solution may be achieved which is exemplified in this thesis in Sections 5.4, 6.7, and 8.4. Therefore, an optimisation problem including all design levels and control needs to be formulated.

2.3 Experimental assessment and comparison of adsorption chillers

As explained in Section 2.2, finding the optimal adsorption-chiller design involves a number of choices on different design levels. To choose the optimal working pair, improve component designs or compare cycle designs, it is necessary to isolate the effect of the specific design choice under consideration on performance. To do so, designs with altered design options have to be assessed and compared. Obviously, it is crucial for the comparison to ensure equal input conditions but, at the same time, it is also crucial to guarantee optimal control for each assessed design. Based on these assessments, designs can be compared, improved, and finally optimised.

Assessment and comparison of adsorption chillers designs can be conducted by experiments or by simulations. In this section, the insights and difficulties arising from experiments on different scales are discussed. In Section 2.4, model-based approaches to compare and optimise adsorption-chiller designs are presented.

The scale of experiments ranges between testing of full-scale adsorption chillers to small-scale experiments for equilibrium data of working pairs. All of these experiments generate insights on different stages of an adsorption chiller. Only results from full-scale adsorption chiller experiments can be evaluated directly in the objective space of an adsorption chiller described by efficiency and power density. Small-scale experiments are only considering parts of an adsorption chiller. Therefore, it is necessary to use assumptions for the not included design levels to estimate the achievable performance by small-scale experiments. In the following, first, results obtained by full-scale experiments are presented. Afterwards, available small-scale experiments are discussed.

2.3.1 Full-scale experiments

Full-scale experiments measure the performance of an adsorption chiller for given input temperatures and volume flow rates in the heat exchangers. Full-scale experiments are useful and necessary to evaluate the performance of a new prototype or a commercially available adsorption chiller.

In literature, large ranges for the performance of adsorption chillers can be found. Figure 2.10 shows the COP/SCP_{ads} pairs obtained by full-scale experiments for 64 adsorption-chiller designs. As can be observed, the range of achieved COP values lies between 0.01 and 0.84, achieved SCP_{ads} values lie in a range of $5.7\,\mathrm{W\,kg^{-1}}$ to $820\,\mathrm{W\,kg^{-1}}$. These large ranges are caused only partly by the adsorption-chiller design itself, but are also caused by design-independent influences, e. g. differences in input conditions or sub-optimal control. Therefore, these values are very difficult to compare. For the purpose of assessing, comparing and improving the intrinsic performance of a design, it is necessary to know and eliminate these design-independent influences on COP and SCP_{ads}.

Input conditions: temperature levels and volume flow rates

One reason for the large ranges of values for COP and SCP_{ads} observed in Figure 2.10 are different input conditions. The input conditions for full-scale experiments are defined by the temperatures and volume flow rates provided at the heat exchanger inlets.

The temperatures (T_{low},T_{mid},T_{high}) provide the driving forces for the adsorption chiller and larger driving forces lead to higher achievable power densities. At the same time, the temperatures define the maximum achievable loading difference and the necessary sensible heat input which both affects efficiency. Figure 2.11 shows the COP/SCP_{ads} pairs colour-coded by the three temperature levels.

Figure 2.11 (a) shows the high temperature level T_{high} in a range from $60\,^{\circ}\mathrm{C}$ to $100\,^{\circ}\mathrm{C}$. It can be observed that higher driving temperatures lead to higher SCP_{ads}

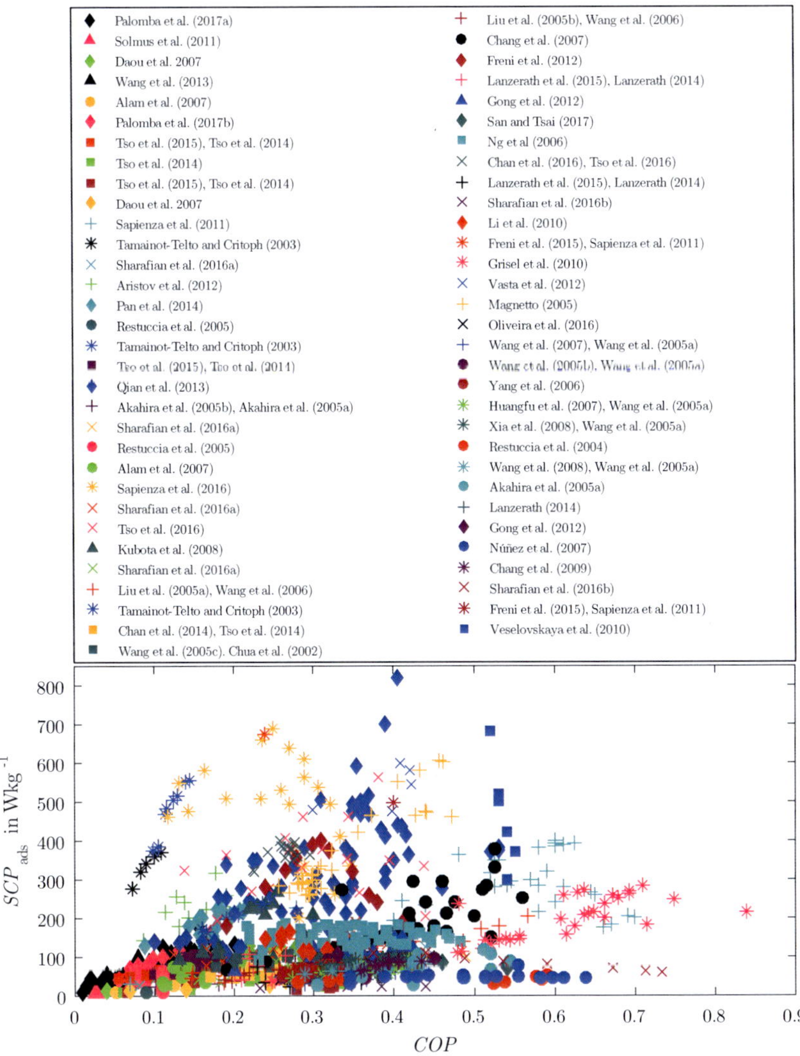

Figure 2.10: COP and SCP_{ads} values obtained by full-scale experiments of adsorption chillers. Different designs are distinguished by colour.

values, but that there is also a very heterogeneous distribution of temperatures around $SCP_{\mathrm{ads}} \approx 100\,\mathrm{W\,kg^{-1}}$.

Figure 2.11 (b) shows the influence of T_{mid} on performance. The picture shows a very heterogeneous distribution of T_{mid} all over the solution space. It is very difficult to see a trend in the data, since the influence of T_{mid} is masked by other design, control and input parameter.

The same problem as for T_{mid} occurs for the effect of the low temperature level T_{low} (Figure 2.11 (c)). A correlation between temperature level and performance can hardly be recognised due to the overlaying effects.

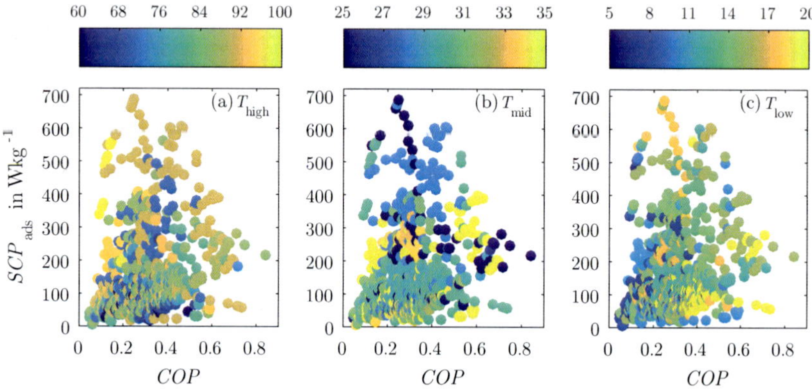

Figure 2.11: COP and SCP_{ads} values obtained by full-scale experiments distinguished by inlet temperatures (for sources see Figure 2.10). The experiments are colour-coded by reported input temperature level. Left: Evaluation for T_{high}. Middle: Evaluation for T_{mid}. Right: Evaluation for T_{low}.

The colour-code of the shown data is based on the set temperatures used for the experiment. This set temperatures have to be distinguished from the time-resolved temperatures provided by the test bench in the actual adsorption cycle. It is very difficult to provide constant input temperatures for a full-scale experiment, since the heating and cooling load of the chiller vary and have to be compensated by the test bench (Vasta et al., 2016). This means that the performance of the adsorption chillers also strongly depends on the test bench and its capability to provide constant temperature inputs.

Finally, also the provided volume flow rates strongly influences the performance. The volume flow rates $\dot{V}_i$ determine the heat capacity flow rate ($\dot{V}_i\, \varrho_i\, c_{\mathrm{p},i}$) in the heat exchangers, where ϱ_i is the density and $c_{p,i}$ is the specific heat capacity of the heat exchanger fluid. The lower the heat capacity flow rate, the stronger is the temperature gradient of the heat exchanger fluid in the heat exchanger. Thus, a difference in heat capacity flow rate leads to a difference in average heat exchanger fluid temperature which directly affects the performance. Chan et al. (2015) varied the volume flow rate of the evaporator $\dot{V}_E = 2 - 7.5\,\mathrm{l\,min^{-1}}$ finding a difference in SCP_{ads} of up to 13 % and in COP of up to 15 %. The effect of the heat capacity flow of each heat exchanger is further analysed in Section 5.3.2.

The discussion on the effect of input conditions on performance shows that it is very difficult to use experimental data from full-scale experiments for a systematic evaluation, comparison, and improvement of design.

Multi-objective design and optimal control

As described in Section 2.1.3, efficiency and power density are two competing objectives, leading to a multi-objective design problem. Thus, there is not a single point where both efficiency and power density reach their maximum value. A higher efficiency comes at the cost of a lower power density and vice versa. Therefore, it is not sufficient to describe the performance of an adsorption-chiller design by a single COP and SCP_{ads} value. Instead, to fully assess a design, a so-called Pareto frontier is needed. A Pareto frontier includes all points which are not dominated by another point in all dimensions of the objective function (Ehrgott, 2010). A Pareto frontier includes the maximum values of each objective and also the trade-off between the objectives. Miles and Shelton (1996) and Metcalf et al. (2012) explicitly use the Pareto frontier to compare heat exchanger and cycle designs of adsorption heat pumps.

For some full-scale experiments reported in literature, it is possible to determine the Pareto frontier for a fixed set of input conditions. Figure 2.12 shows the Pareto frontiers for all experiments where 3 or more data points are available for one set of inlet temperatures and volume flow rates. Additionally, the non-Pareto-optimal points are shown to illustrate the experimental effort necessary to obtain the Pareto frontier.

Since some designs promote a high efficiency and some designs promote a high power density, Pareto frontiers for these designs cross at certain COP/SCP_{ads} pairs (e. g. Grisel et al. (2010) and Sapienza et al. (2011)). Which design is favourable depends on the designer's preferences and on the application (Leong and Liu, 2004b): For applications utilising waste heat, power density is more important because the cost of energy input is almost zero. For applications with a limited heat source, efficiency is more important than power density.

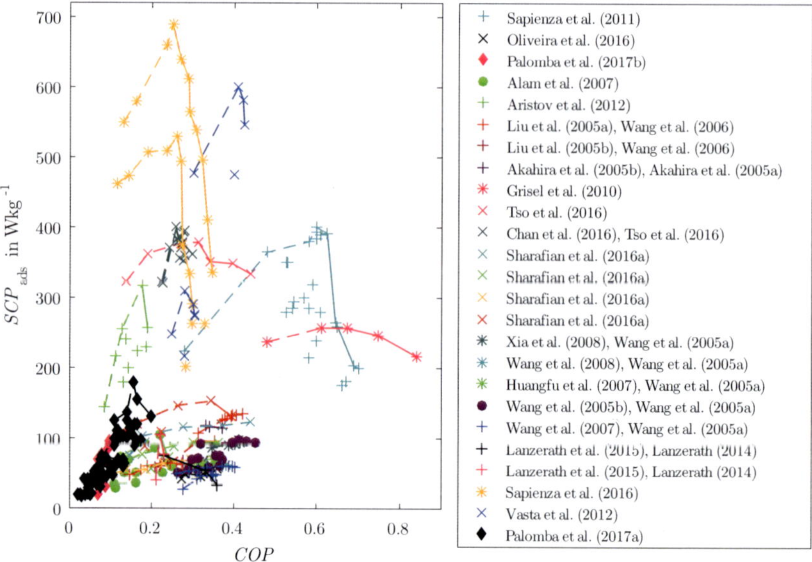

Figure 2.12: Pareto frontiers of reported full-scale experiments with more than 3 or more data points for a single set of input conditions (temperatures and volume flow rates).

Alam et al. (2003) propose to combine efficiency and power density using a weighted profit function based on the prices of input and output heat. Thereby, the multi-objective design problem would be reduced to a single objective problem. In the reduced problem, the user's preferences are incorporated implicitly by the prices and the explicit information on efficiency and power density is lost. For the purpose of comparison without user's preferences, the full Pareto frontier should be determined.

For a given adsorption-chiller design and given input conditions, the Pareto frontier depends only on the control. Still all control parameters should be optimised with respect to all objectives (Pons, 1996; Rezk and Al-Dadah, 2012), leading to a high experimental effort.

Even if the full Pareto frontier is determined and the same input conditions could be ensured, designs in full-scale experiments may vary in a number of choices at the same time. Thus, the differences in performance can hardly be ascribed to a specific design choice. Therefore, small-scale experiments have been developed, analysing only parts of an adsorption chiller.

2.3.2 Small-scale experiments

Small-scale experiments aim at getting insights on specific design levels of an adsorption chiller. Compared to full-scale experiments, it is possible to separate the effects of certain design measures regarding equilibrium data, and heat and mass transfer characteristics. The smaller the scale of the experiment, the more specific the obtained information. At the same time, it is more difficult to estimate the effect on efficiency and power density of the chiller, since the necessary assumptions on other design levels and control increase. Nevertheless, it is helpful and common practice to estimate COP and SCP_{ads} based on small-scale experiments (Aristov et al., 2008; Aristov, 2014a; Sapienza et al., 2014). In the following, the most common small-scale experiments are presented and the necessary assumptions to derive the performance indicators of an adsorption chiller are named. The experiments are presented going from "large" to "small".

Lanzerath et al. (2014) characterise the components of an adsorption chiller using a modular one-bed cyclic test bench. This test bench consists of a full-scale evaporator, condenser, and adsorber. Compared to a full-scale prototype, the change of components is easily possible due to the modularity of the test bench. With this test bench, only the simple adsorption cycle can be experimentally investigated by measuring the heat flow rates at evaporator, condenser, and adsorber. To estimate COP and SCP_{ads} of an advanced cycle, additional assumptions on cycle design would be necessary.

The gravimetric large-temperature-jump (G-LTJ) experiment is designed to investigate the kinetics of small-scale adsorbers (Wittstadt et al., 2008; Sapienza et al., 2014). To do so, the adsorber is exposed to an isobaric temperature-jump and the mass of adsorbed water is measured by a scale. Compared to a modular one-bed cyclic test bench, a huge evaporator and condenser are used to isolate the effect of the adsorber design. The G-LTJ experiment characterises the adsorber, i. e. kinetics of working pair, adsorbent configuration, and adsorber heat exchanger. The gravimetric large-temperature-jump does not measure a full cycle. Therefore, assumptions on evaporator, condenser, and control are necessary to estimate the achievable COP and SCP_{ads}.

The large-temperature-jump (LTJ) experiment measures the kinetics of a small probe of adsorbent given a fast quasi-isobaric temperature jump (Aristov et al., 2008). The LTJ is used to determine kinetics of working pairs and adsorbent configurations, such as pellet sizes or coatings. Often, the characteristic time, which is needed to achieve a certain percentage of loading difference, is compared (Aristov et al., 2012a). This characteristic time is also used to determine a maximum power density: the achieved loading difference is divided by the characteristic time and the mass of adsorbent. Since the LTJ experiment measures small probes on a plate which is perfectly heated and cooled, the estimation of power density does not include most heat and mass transfer resistances discussed in Section 2.2.2.

The thermo-gravimetrical scale allows to determine the working pair's equilibrium data by exposing a small probe of adsorbent to a defined adsorbate temperature and pressure and measuring loading using a high-resolution scale (Núñez, 2001). To estimate a COP and SCP_{ads} from this data, assumptions on all other levels are necessary.

Figure 2.13 links the small-scale experiments and the obtained information with the design levels. In general, a trade-off between the specificness of the insights gained

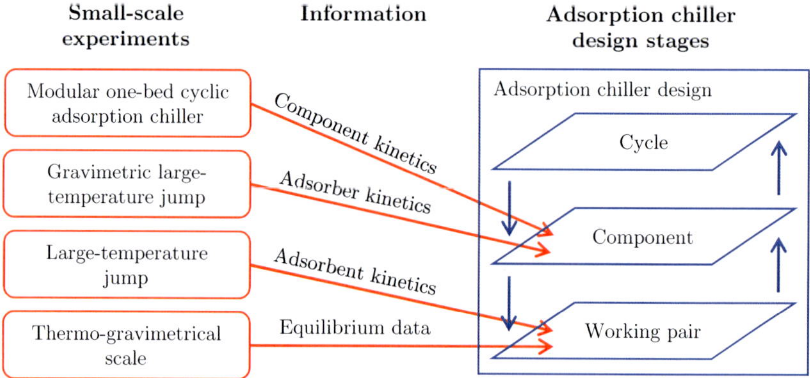

Figure 2.13: Information obtained by small-scale experiments for characterisation and link to adsorption-chiller design levels.

by an experiment and the necessary assumptions to estimate COP and SCP_{ads} of an adsorption chiller exists, e. g. the thermo-gravimetrical scale gives very detailed information on the equilibrium data of a working pair, but COP and SCP_{ads} can only be estimated roughly. A way to overcome this trade-off is the combination of model-based assessment and experimental calibration as proposed by our working group (Graf et al., 2016). The described small-scale experiments are used to determine necessary model parameters such as equilibrium data, and heat and mass transfer coefficients. A simulation model is used to predict the performance of a full-scale chiller including explicit assumptions for all design levels. Compared to experiments, a simulation model also allows to guarantee equal input conditions and to determine the optimal control more quickly.

In the next section, an overview on available simulation models is given. Additionally, optimisation approaches are presented that are used to find the optimal control of adsorption chillers.

2.4 Simulation and optimisation of adsorption chillers

The challenges in designing, assessing, and comparing adsorption chillers presented in Sections 2.2 and 2.3 set high requirements for experiments:

First, the manifold degrees of freedom in design (cf. Section 2.2) lead to a large amount of possible combinations. Thus, experimental investigation of all design parameters is very expensive and time consuming. To reduce these costs, simulation models are used for fast exploring of the effects of design parameters.

Second, it is very difficult to compare full-scale experiments conducted at different test benches, due to the large influence of input conditions and control optimisation (cf. Section 2.3.1). Riffel et al. (2010) conclude that it is necessary to use simulation models to guarantee equal conditions for assessment.

In this thesis, a simulation-based approach of assessing and designing adsorption chillers is developed. The proposed approach allows to simultaneously optimise design and control of adsorption chillers on all levels. Experimental data are incorporated to achieve valid results.

In this section, first, available simulation approaches of different granularity are reviewed. From this discussion, a model granularity is identified which allows to simultaneously incorporate the effects of all design and control levels. Second, available calibration and validation of simulation models in literature are summarised. Third, optimisation techniques are discussed and evaluated, which are used to find the optimal control of adsorption chillers.

2.4.1 Simulation of adsorption chillers

The simulation-based approach proposed in this thesis sets two competing requirements to a simulation model: (1) The simulation model has to be complex enough to consider the key design parameters on each design stage. These parameters are the equilibrium data of the working pair, the heat and mass transfer characteristics of adsorber, evaporator and condenser, the effect of enhanced cycle designs, and the effect of control. (2) The simulation model has to be fast enough to allow for optimisation of control and design choices on adsorption chiller system level.

Reviews of Pesaran et al. (2016); Teng et al. (2016) and Yong and Sumathy (2002) divide adsorption chiller models into three classes: Thermodynamic models, lumped-parameter models, and distributed-parameter models. In the following, a short overview on available studies and investigated parameters is given. For more information, the author refers to the mentioned reviews.

Thermodynamic models

Thermodynamic models consider the equilibrium states for a given temperature triple. By including an overall energy and entropy balance, the achievable maximum efficiency for a certain working pair can be estimated. Cacciola and Restuccia (1995) used this model class to compare working pairs based on their maximum efficiency.

Meunier et al. (1998) compared adsorption-chiller cycles and compression chillers by using equivalent Carnot cycles. Pons and Poyelle (1999) and Alam et al. (2004) also used thermodynamic models to compare cycle configurations including active heat recovery, thermal wave and mass recovery.

Since thermodynamic models do not include system dynamics, it is not possible to estimate the power density of a system. Therefore, thermodynamic models are considered as a starting point which determines an upper bound regarding efficiency (cf. Section 8.2).

Lumped-parameter models

Going beyond thermodynamic models, lumped-parameter models also consider heat and/or mass transfer limitations by overall transfer coefficients. Lumped models assume uniform temperature, pressure, and loading in each component. These models are widely used for estimation of efficiency and power density of adsorption chillers for varying temperature levels, input conditions, component characteristics (heat and mass transfer coefficients, thermal mass), or control.

The first dynamic lumped-parameter model for an adsorption chiller was proposed by Douss et al. (1988). They investigated transient temperature profiles and influence of heat transfer coefficient on COP. Saha et al. (1995) and Gong et al. (2011) used lumped models to investigate the influence of temperature levels. The effect of volume flow rates is investigated by Pons (1996) and Voyiatzis et al. (2008).

Next to the input conditions, also component characteristics are investigated. Cho and Kim (1992) varied the heat transfer coefficients and investigated the effect on performance. Hamamoto et al. (2005) changed the ratio of adsorbent beds in a multi-bed configuration.

Pons and Feng (1997) used lumped-parameter models to investigate cycle designs. They compared active heat recovery and a thermal wave cycle for a two-bed configuration. Chua et al. (2001) compared 2/4/6-bed multi-stage systems and Rezk and Al-Dadah (2012) investigated the effect of heat and mass recovery for a two-bed system.

We used a validated 1-d discretised lumped-parameter model to investigate the effect of component sizing on the volumetric SCP (Lanzerath et al., 2015). In this model, we disctretised the heat exchangers of evaporator, condenser, and adsorber in flow direction (1-d discretisation) leading to a very good agreement of outlet temperatures of simulation and experiment.

Finally, also the effect of control on performance is investigated using lumped parameter models by Chua et al. (1999), Nasruddin (2005), and Sharafian and Bahrami (2015).

This short overview shows that lumped models are well suited and widely used to investigate effects of design, control, and input conditions on the performance of adsorption chillers. Nevertheless, lumped models do not include spatial geometric information which limits their capability to optimise component geometries.

Distributed-parameter models

The third class are spatially discretised models (distributed-parameter models). Compared to lumped-parameter models, this model class allows for optimisation of component geometries.

Zhang and Wang (1999) investigated the influence of fin number, fin height, and thermal conductivity for a finned tube adsorber on COP and cooling power. The study assumed an ideal evaporator and condenser, a simple cycle and does not provide information on how phase times are obtained.

Leong and Liu (2004a) analysed the effect of adsorbent-bed thickness, adsorbent-particle size, bed porosity and driving temperature T_{high} on COP and SCP_{ads}. In accordance with Zhang and Wang (1999), ideal evaporator and condenser and a simple cycle are assumed.

Freni et al. (2012a) simulated the kinetics of loose adsorbent grains on a metal plate. They simulated a large-temperature jump experiment and investigated the effects of grain number on temperature distribution and water uptake. Their model can be used to find heat and mass transfer coefficients depending on adsorber configuration and grain size by comparing simulation and experiment, but the model is too complex for system simulation: simulation of an adsorption/desorption phase for only four grains took up to 10 min.

Figure 2.14 summarises the model classes, their field of application and the trade-off between calculation speed and complexity. The short overview shows that many aspects of adsorption chillers and heat pumps have been studied. This has been done by varying certain design or operating aspects based on a reference case and exploring the effect on performance. Pesaran et al. (2016) conclude in their review that "a system-level study that includes the simultaneous optimisation of the adsorber bed specifications and operational parameters of the system is lacking."

The aim of this thesis is to conduct such a simultaneous optimisation of design and control on adsorption chiller level. Additionally, the optimisation is conducted regarding multiple objectives (2 in this thesis) leading to multiple simultaneous optimisations to exploit the full Pareto frontier. The class of dynamic lumped-parameter models with 1-d disctretised heat exchangers is chosen to include the trade-off between efficiency and power density, but also to enable optimisation of all design levels under optimal control.

2.4.2 Optimisation of adsorption chillers

Mathematical optimisation came up in adsorption chiller literature only in recent years (Gräber et al., 2011; Rezk and Al-Dadah, 2012; Rahman et al., 2013). So far, optimisation for adsorption chillers is used to determine optimal phase times. For design improvement, sensitivity analysis is commonly used (cf. Section 2.4.1). Thus, only one parameter is varied at a time. This type of analysis allows to investigate the effect of a design choice, but does not include interactions between design choices and also does not lead to an optimal design. Thus, it would be desirable to enhance optimisation from control only to control and design.

The optimisation approaches used to find the optimal control for adsorption chillers can be classified into 3 categories: full factorial design, derivative-free optimisation (Jones et al., 2002), and derivative-based optimisation (Nocedal and Wright, 2006).

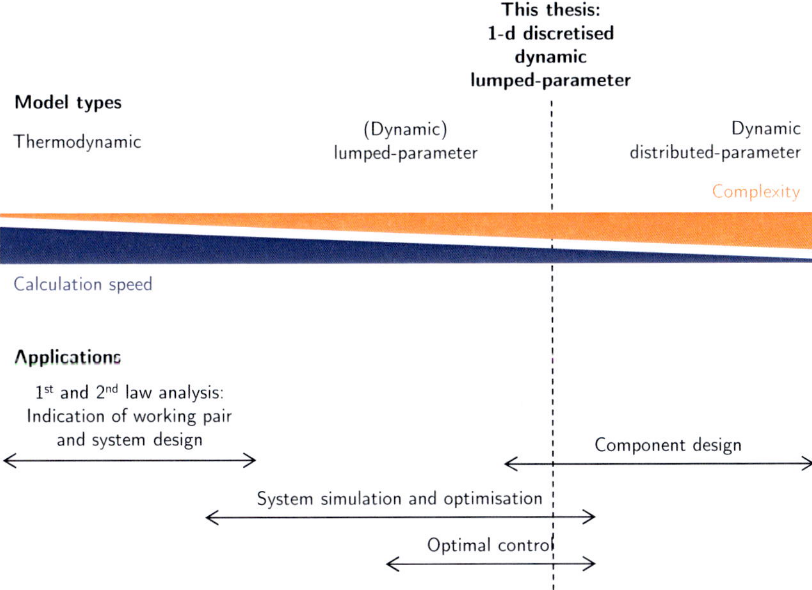

Figure 2.14: Model classes, their field of applications and the trade-off between calculation speed and model complexity.

Full factorial design

All possible combinations of parameters are evaluated within a full factorial design. An advantage of this approach is that the entire solution space is investigated which ensures to find the global optimum. Of course this approach is very time consuming and the optimisation time increases exponentially with number of optimisation parameters. Zajaczkowski (2016) used this approach to find the optimal switching time between adsorption and desorption and the optimal time for mass recovery, but without optimising the overall cycle time. Critoph et al. (2016) used a full factorial design to find the optimal times of adsorption and desorption for a heat pump application.

Derivative-free optimisation

Derivative-free or heuristic optimisation algorithms use only the information provided by the objective function but no derivative information. Based on the evaluation of a starting population, derivative-free optimisation uses heuristics to modify the population. Popular derivative free algorithms are the genetic algorithm (Holland, 1975) or the particle swarm optimisation algorithm (Kennedy and Russell, 1995) which have also been used to find optimal control parameters for adsorption chillers. Derivative-free optimisations are very robust, since they do not use derivatives: They can cope also with non-smooth or noisy objective functions. On the other hand, derivative-free optimisation techniques are less efficient than derivative-based algorithms. Körkel et al. (2005) showed that for a system of differential algebraic equations a derivative-based approach was 37.5 faster than a derivative-free optimisation algorithm.

Rezk and Al-Dadah (2012) and El-Sharkawy et al. (2013) both used genetic algorithms to determine the optimal phase times of two-bed adsorption chillers. Miyazaki et al. used the particle swarm optimisation algorithm to determine the optimal phase times for a two-bed (Miyazaki and Akisawa, 2009) and a three-bed adsorption chiller (Rahman et al., 2013).

Derivative-based optimisation

Derivative-based optimisation algorithms use the first and second derivatives with respect to the parameters and constraints to find an optimal solution. Derivative-based optimisation algorithms are potentially more efficient than derivative-free optimisation algorithms (Körkel et al., 2005). At the same time, they pose higher requirements regarding the objective function, since it has to be differentiable with respect to all parameters and constraints.

Derivative-based optimisation is not commonly used for optimisation of adsorption energy systems. Only Gräber et al. (2011) use an efficient multiple-shooting algorithm to find the optimal phase times for a two-bed chiller. In this thesis, the same approach is used. Additionally to the phase times, also design parameters are included, leading to the simultaneous optimisation of design and control.

2.5 Contribution of this thesis: from simulation to optimisation of adsorption chillers

Simulation models are well established and widely used to investigate the performance of adsorption chillers. The effect of design choices is commonly examined by sensitivity analyses without changing other design or control parameters. Thus, designs may be compared at poor operating points and interactions between design choices are not considered. To overcome these limitations, two major requirements can be derived:

1. It is necessary to re-optimise control when changing a design parameter. This re-optimisation ensures that only the intrinsic effect of the design choice and not the effect of poor control is investigated.
2. A simultaneous optimisation of design choices is desirable to include interactions.

The required optimisations have to be carried out for dynamic systems and for at least two objectives. In this thesis, a 1-d discretised dynamic lumped-parameter model is used, which is formally described by a set of differential and algebraic equations (DAE). As objectives, SCP and COP are considered. Thus, a multi-objective DAE-constrained optimal control problem arises. In this thesis, the resulting problem is formulated and solved.

The main contributions of this thesis can be clustered in three parts, which are illustrated in Figure 2.15:

Object-oriented, dynamic-model library for adsorption energy systems (Part I)

In Chapter 3, an object-oriented model library for adsorption energy systems in the modelling language "Modelica" is developed. The library allows to quickly develop dynamic models of adsorption energy systems. Thus, fast modelling of new cycle designs is enabled. The developed library is also suitable for adsorption heat pumps and desiccant systems.

Multi-objective optimal design and control (Part II)

A framework for multi-objective optimal design and control of adsorption chillers is developed. The framework pursues the overall goal of a simultaneous optimisation of all design levels, working pair, component design, and cycle design while ensuring optimal control. On the way towards simultaneous optimisation of all design levels, the framework is used to assess, compare, and optimise designs on different design levels while ensuring optimal control.

In Chapter 4, the framework is presented and a mixed-integer optimisation problem is formulated including all design levels. This optimisation problem represents the overall goal in designing adsorption chillers. In the Chapters 5-8, this goal is approached stepwise by successively enhancing the design levels considered in the optimisation problem.

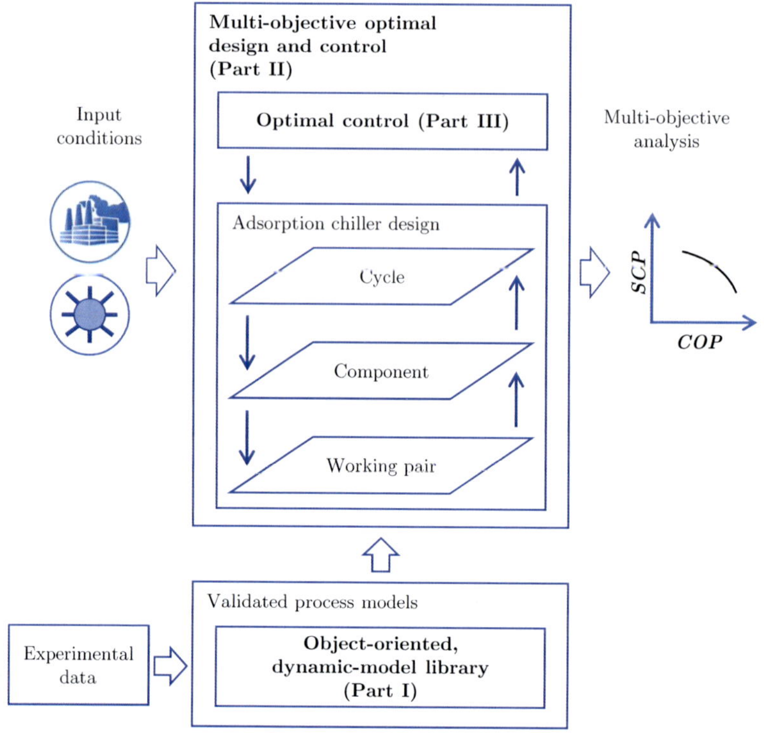

Figure 2.15: Main contributions of this thesis: model-based, multi-objective optimal design and control of adsorption energy systems based on an object-oriented, dynamic-model library for adsorption systems.

In Chapter 5, the framework is exemplified by optimising the grain size and fin number of a finned-tube adsorber with silica-gel grains. For the cycle design, a one-bed chiller configuration is used. Evaporator and condenser are modelled as ideal components, providing a constant pressure. Also in Chapter 5, the used optimisation algorithm is explained in detail.

In Chapter 6, the example is enhanced by including models for a real evaporator and condenser. The resulting optimisation problem simultaneously optimises adsorber design, evaporator and condenser sizing, and control.

In Chapter 7, the framework is used to investigate the influence of the working pair on optimal adsorber design, component sizing, and control.

Finally, in Chapter 8, five two-bed chiller cycle designs are compared using the Pareto frontier for SCP and COP. The multi-objective optimisation reveals that it is beneficial to change the operation mode of a two-bed chiller when moving from high efficiency to high power density.

Optimal control during operation: model predictive control and new heuristics (Part III)

In Chapter 9, a closed-loop model predictive control (MPC) for a one-bed adsorption chiller is presented. The MPC uses the model library and the optimisation techniques already employed for the simultaneous design and control optimisation. The model predictive control is tested experimentally for the first time. Additionally, new heat-flow-based heuristics for a (near) Pareto-optimal operation of one-bed chillers are developed and compared to the model predictive control.

Part I

Modelling of adsorption energy systems

Chapter 3

Adsorption Energy Systems Library

The design of adsorption systems is challenging due to their intrinsic dynamic nature: During operation, adsorption devices switch between ad- and desorption phases, and, thus, they work discontinuously (cf. Section 2.1). Besides, adsorption devices consist of several components, all being influencing the performance. In an optimal design, these components need to be balanced to avoid oversized components on the one hand and bottlenecks on the other hand (cf. Section 2.2). To meet this optimal design challenge, dynamic models have been developed to describe and improve adsorption-based devices (cf. Section 2.4). These models have been developed mainly to study the performance of one specific configuration. The model library presented in this chapter follows a generic modular approach to describe all kinds of adsorption-based thermal devices.

Contents of this chapter have been published in:

Bau, U., Lanzerath, F., Gräber, M., Schreiber, H., Thielen, N., and Bardow, A. (2014). Adsorption energy systems library - Modeling adsorption based chillers, heat pumps, thermal storages and desiccant systems. In Tummescheit, H. and Arzén, K.-E., editors, *Proceedings of the 10th International Modelica Conference*, Linköping electronic conference proceedings, pages 875-883, Linköping. Modelica Association.

Schicktanz and Núñez (2009) were the first who described an adsorption chiller by using a modular approach in Modelica, but they did not report the development of a generic library. Joos et al. (2009) presented a modular Modelica library for separation processes including adsorption. This library can be used to model separation processes, but is not designed for thermal applications, in contrast to the presented library in this thesis.

In this chapter, the Adsorption Energy Systems Library is presented, following the library structure in Figure 3.1. First, in Section 3.1, the implementation of the working pair's equilibrium data in the media model is shown. Afterwards, in Section 3.2, the main cell models are presented: adsorbent, vapour-liquid phase separator, liquid cell, and gas cell. These cell models are volumes, which include the media data, an energy balance, and a mass balance. The cells can be connected by heat and mass transfer models (Section 3.3). Finally, in Section 3.4, three validated examples are presented, which illustrate the flexibility of the library.

3.1 Media

As described in Section 2.2.1, the core of every adsorption-based device is the working pair. The media model describes the characteristic properties of the working pairs. All properties are described by two independent variables (cf. Section 2.1.1). For a given temperature T and loading w, the model returns the equilibrium pressure p_{ad}, internal energy of the adsorbed fluid u_{ad}, and specific heat capacity of both adsorbent c_{sor} and adsorbate c_{ad}. Temperature and loading are used as independent variables for the media model, since these two variables are used as differential states in the adsorbent model.

In principle, every adsorption model can be implemented. The working pairs used in this thesis are described by the model of Dubinin (1967), with experimentally determined characteristic functions by Núñez (2001) and Goldsworthy (2014). Additionally, the library contains a large number of equilibrium data reported in literature, allowing for a fast exchange of the working pair.

For the implementation of the Dubinin model, the equilibrium pressure p_{ad} is calculated as:

$$p_{\text{ad}} = p_{\text{s}}(T) \exp\left(-\frac{A}{RT}\right), \tag{3.1}$$

with

$$A = W(A)^{-1} \tag{3.2}$$

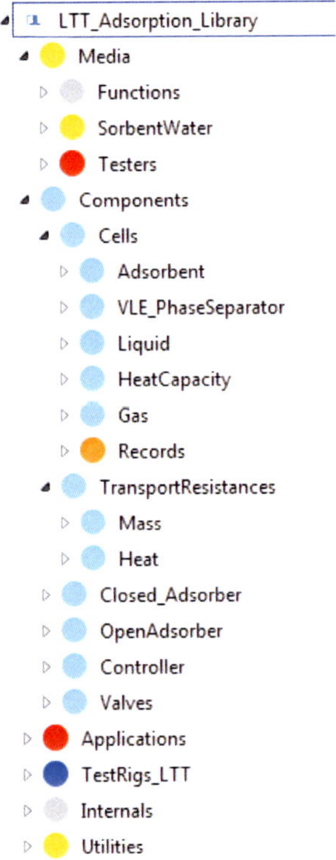

Figure 3.1: Structure of the LTT Adsorption Energy Systems Library.

as the inverse characteristic curve (cf. Section 2.1.1). The characteristic curves used in this thesis can be found in Appendix A.

The specific internal energy is calculated as

$$u_{\mathrm{ad}} = h_{\mathrm{ad}} - \frac{p_{\mathrm{ad}}}{\varrho_{\mathrm{ad}}(T)} , \tag{3.3}$$

with

$$h_{\mathrm{ad}} = h_{\mathrm{l}} - \Delta h_{\mathrm{bond}} . \tag{3.4}$$

The specific bonding enthalpy Δh_{bond} is calculated by Equation (2.10). The specific internal energy u_{ad} and the specific enthalpy h_{ad} calculated by the media model are marginal values, i. e. they describe the specific values of the last adsorbed molecule at

w and T. In contrast to the marginal specific internal energy u_{ad}, the average specific internal energy of the adsorbate bulk-phase is denoted $\overline{u}_{\text{ad}}$.

The total heat capacity of sorbent and adsorbate c_{tot} is the sum of the individual heat capacities:

$$c_{\text{tot}} = c_{\text{sor}} + w\, c_{\text{ad}}(T)\ , \tag{3.5}$$

where c_{sor} is assumed constant for a specific adsorbent (cf. Appendix A).

For the adsorbate density $\varrho_{\text{ad}}(T)$, specific heat capacity $c_{\text{ad}}(T)$ and thermal coefficient of expansion $\alpha(T)$, the properties of the liquid phase are used.

In the current implementation of the library, all fluid properties (vapour, liquid and two-phase region) are based on TILMedia (Gräber et al., 2010), a fluid property library provided by TLK-Thermo GmbH. TILMedia provides fluid properties implementation for Modelica which is optimised regarding simulation speed. The fluid properties itself are based on the National Institute of Standards and Technology (NIST) Standard Reference Database (Lemmon et al., 2013).

3.2 Cells

The following components are the smallest units of the library. All extended components and device models are based on these basic components. The cells are volume elements, which include a media model, an energy balance, and a mass balance. The cells have so called connectors, which allow for heat and mass exchange with other cells. A spatial discretisation is realised by connecting cells following the finite volume approach.

3.2.1 Adsorbent

In the adsorbent cell, the media model of the working pair is used. The differential states of the adsorbent are the temperature T and the loading w. In the adsorbent cell, it is assumed that the entire mass of adsorbate is in adsorbed state, since the mass in the vapour phase at low pressures is very low compared to the adsorbed mass. The mass balance of the adsorbent reads

$$\frac{\mathrm{d}w}{\mathrm{d}t} = \frac{1}{m_{\text{sor}}} \left(\dot{m}_{\text{fl,in}} - \dot{m}_{\text{fl,out}}\right)\ . \tag{3.6}$$

The energy balance of the adsorbent reads

$$\frac{\mathrm{d}U_{\text{sor+ad}}(w,T)}{\mathrm{d}t} = \dot{m}_{\text{fl,in}}\, h_{\text{in}} - \dot{m}_{\text{fl,out}}\, h_{\text{out}}(T) + \dot{Q} \tag{3.7}$$

$$\frac{\mathrm{d}\left(u_{\mathrm{sor}}(T)+w\,\overline{u}_{\mathrm{ad}}(w,T)\right)}{\mathrm{d}t}=\frac{1}{m_{\mathrm{sor}}}\left(\dot{m}_{\mathrm{fl,in}}\,h_{\mathrm{in}}-\dot{m}_{\mathrm{fl,out}}\,h_{\mathrm{out}}(T)+\dot{Q}\right), \tag{3.8}$$

where $U_{\mathrm{sor+ad}}(w,T)$ is the internal energy of adsorbent and adsorbate which can be expressed by the specific internal energy of the adsorbent $u_{\mathrm{sor}}(T)$ and the specific internal energy of the adsorbate in adsorbed state $\overline{u}_{\mathrm{ad}}(w,T)$. This specific average internal energy is not to confuse with the marginal specific internal energy of the last adsorbed molecule u_{ad} (cf. Equation (3.3)). On the right side, $\dot{m}_{\mathrm{fl,in}}\,h_{\mathrm{in}}$ is the entering enthalpy flow, $\dot{m}_{\mathrm{fl,out}}\,h_{\mathrm{out}}$ is the leaving enthalpy flow, and $\dot{Q}$ describes the heat flow which enters or leaves the adsorbent cell. The specific enthalpy of the entering mass flow h_{in} is determined by the connected cell where the mass flow is coming from. The specific enthalpy of the leaving mass flow h_{out} is assumed to be the enthalpy of adsorbate in the state of saturated vapour $h_s^v(T)$.

To express the left side of Equation (3.8) in terms of temperature and loading, the equation can be rewritten as

$$\frac{\partial\left(u_{\mathrm{sor}}(T)+w\,\overline{u}_{\mathrm{ad}}(w,T)\right)}{\partial T}\frac{\mathrm{d}T}{\mathrm{d}t}+\frac{\partial\left(u_{\mathrm{sor}}(T)+w\,\overline{u}_{\mathrm{ad}}(w,T)\right)}{\partial w}\frac{\mathrm{d}w}{\mathrm{d}t} \tag{3.9}$$

$$=\left(c_{\mathrm{sor}}+wc_{\mathrm{ad}}\right)\frac{\mathrm{d}T}{\mathrm{d}t}+\frac{\partial\left(w\,\overline{u}_{\mathrm{ad}}(w,T)\right)}{\partial w}\frac{\mathrm{d}w}{\mathrm{d}t} \tag{3.10}$$

$$=\left(c_{\mathrm{sor}}+wc_{\mathrm{ad}}\right)\frac{\mathrm{d}T}{\mathrm{d}t}+u_{\mathrm{ad}}(w,T)\frac{\mathrm{d}w}{\mathrm{d}t} \tag{3.11}$$

$$=\frac{1}{m_{\mathrm{sor}}}\left(\dot{m}_{\mathrm{fl,in}}\,h_{\mathrm{in}}-\dot{m}_{\mathrm{fl,out}}\,h_{\mathrm{out}}(T)+\dot{Q}\right), \tag{3.12}$$

by using the definition of $\overline{u}_{\mathrm{ad}}=\frac{1}{w}\int_0^w u_{\mathrm{ad}}(w,T)\mathrm{d}w$:

$$\frac{\partial\left(w\,\overline{u}_{\mathrm{ad}}(w,T)\right)}{\partial w}=\frac{\partial\left(w\,\frac{1}{w}\int_0^w u_{\mathrm{ad}}(w,T)\mathrm{d}w\right)}{\partial w}=u_{\mathrm{ad}}(w,T)\,. \tag{3.13}$$

The adsorbent model contains heat and fluid ports, allowing for heat and mass transfer. The fluid ports are used to connect the adsorbent cell to the working fluid, the heat port is used to connect the adsorbent cell to the heat exchanger. Each port contains a potential variable and a flow variable. For the fluid port, the potential variable is the pressure p_{ad} and the flow variable is the mass flow $\dot{m}$. For the heat port, the potential variable is the temperature T and the flow variable is the heat flow rate $\dot{Q}$. For more information about the concept of ports in Modelica see Fritzson (2015). Figure 3.2 shows the ports and differential states of the adsorbent cell.

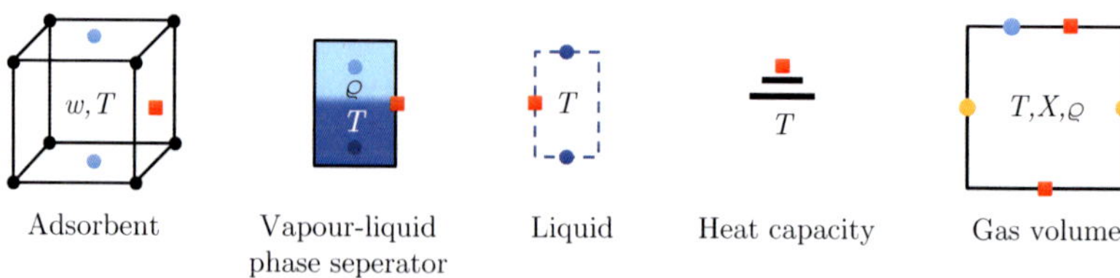

Figure 3.2: Scheme of cell components including the differential state variables and ports (red: heat ports, blue: fluid ports, yellow: gas ports).

3.2.2 Vapour-liquid phase separator

The vapour-liquid phase separator cell is used to model the working fluid in the evaporator and condenser. It assumes that the fluid is in the two phase region. As media model, a vapour-liquid equilibrium model of the adsorbate is used. The differential states of the vapour-liquid phase separator are the temperature T and the density ϱ. The mass balance reads

$$V\frac{\mathrm{d}\varrho}{\mathrm{d}t} = \dot{m}_{\mathrm{fl,in}} - \dot{m}_{\mathrm{fl,out}} \,. \tag{3.14}$$

The energy balance reads

$$\left(u + \varrho\left.\frac{\partial u}{\partial \varrho}\right|_T\right)\frac{\mathrm{d}\varrho}{\mathrm{d}t} + \varrho\, c_{\mathrm{v}}\frac{\mathrm{d}T}{\mathrm{d}t} = \frac{1}{V}\left(\dot{m}_{\mathrm{fl,in}}\, h_{\mathrm{in}} - \dot{m}_{\mathrm{fl,out}}\, h_{\mathrm{out}}(T) + \dot{Q}\right) \,, \tag{3.15}$$

where V is the volume of the cell and $c_{\mathrm{v}} = \left.\frac{\partial u}{\partial T}\right|_\varrho$ is the isochoric heat capacity. The energy balance is explicitly written in temperature and density to increase simulation speed. The partial derivative of the internal energy with respect to the density can be expressed as (Tummescheit, 2002):

$$\left.\frac{\partial u}{\partial \varrho}\right|_T = \frac{1}{\varrho^2}\left(-p + T\frac{\mathrm{d}p}{\mathrm{d}T}\right) \,. \tag{3.16}$$

The vapour-liquid phase separator model also contains heat and fluid ports (see Figure 3.2). The pressure which is used as potential variable for the fluid ports is calculated by the saturation pressure

$$p = p_{\mathrm{s}}(T). \tag{3.17}$$

3.2.3 Liquid

The liquid cell is used to describe the heat exchanger fluid in the heat exchangers. The used media model describes an incompressible fluid. Therefore, the mass balance reads:

$$0 = \dot{m}_{\mathrm{fl,in}} - \dot{m}_{\mathrm{fl,out}} \tag{3.18}$$

and the energy balance reads:

$$c(T)\frac{\mathrm{d}T}{\mathrm{d}t} = \frac{1}{m_{\mathrm{fl}}}\left(\dot{m}_{\mathrm{fl,in}}\, h_{\mathrm{in}} - \dot{m}_{\mathrm{fl,out}}\, h_{\mathrm{out}}(T) + \dot{Q}\right)\,, \tag{3.19}$$

where c is the specific enthalpy of the fluid.

The liquid cell also contains heat and fluid ports which are illustrated in Figure 3.2.

3.2.4 Heat capacity

The heat capacity cell is used to describe solid parts. Therefore, the model has constant mass and no incoming and outgoing mass flows. The energy balance reads:

$$mc\frac{\mathrm{d}T}{\mathrm{d}t} = \dot{Q}\,, \tag{3.20}$$

The heat capacity model only contains a heat port. Figure 3.2 illustrates the port and the differential state of the heat capacity.

3.2.5 Gas volume

Besides adsorption chillers and heat pumps, the Adsorption Energy Systems Library also allows to model desiccant systems. In contrast to an adsorption chiller, a desiccant system is not connected to the pure adsorbate at vacuum conditions. Instead, a desiccant system is connected to a gas mixture of which the adsorbate is one component, e. g. moist air. The adsorbent then exchanges adsorbate and heat with the moist air.

For the case of moist air, the differential states of the gas cell are the gas temperature T, the mass fraction of the adsorbate (water) X, and the air density ϱ_{air}.

The total mass balance reads:

$$\frac{\mathrm{d}\varrho_{\mathrm{air}}}{\mathrm{d}t} = \frac{1}{V}\left(\dot{m}_{\mathrm{air,in}} - \dot{m}_{\mathrm{air,out}} - \dot{m}_{\mathrm{ads}}\right)\,. \tag{3.21}$$

where $\dot{m}_{\text{ads}}$ is the amount of adsorbed water. The mass balance of water is correspondingly:

$$\frac{\mathrm{d}\,(\varrho_{\text{air}}\,X)}{\mathrm{d}t} = \frac{1}{V}\,(\dot{m}_{\text{air,in}} X_{\text{in}} - \dot{m}_{\text{air,out}} X - \dot{m}_{\text{ads}}) \ , \tag{3.22}$$

where X is the water mass fraction

$$X = \frac{m_{\text{water}}}{m_{\text{air}}} \ . \tag{3.23}$$

The energy balance of the gas volume is given by:

$$\frac{\mathrm{d}\,(\varrho_{\text{air}}\,u_{\text{air}})}{\mathrm{d}t} = \frac{1}{V}\left(\dot{m}_{\text{air,in}}\,h_{\text{air,in}} - \dot{m}_{\text{air,out}}\,h_{\text{air,out}} - \dot{m}_{\text{ads}}\,h_{\text{ads}} + \dot{Q}\right) \ . \tag{3.24}$$

where the enthalpy of water being adsorbed h_{ads} is the enthalpy of water vapour at gas temperature T and partial pressure p_{water}.

To allow for a gas flow through the volume, the model has two gas ports. It also contains a heat port and a fluid port which are connected to the adsorbent. The pressure used as potential variable at the fluid port is the partial pressure of water p_{water}. Figure 3.2 illustrates the ports and the differential states of the gas cell.

3.3 Heat and mass transfer

In Section 3.2, the cell models are described. All cell models are volume elements with ports. These ports are used to connect cell models by a transfer model. Based on the potential variable of the ports, this transfer model determines the flow variable. The transfer models do not contain a volume, i. e. they cannot store mass or energy. In the Adsorption Energy Systems Library, several heat and mass transfer models are available.

3.3.1 Heat transfer

The heat transfer model thermally connects two cell models, e. g. the adsorbent cell to a heat capacity cell or to a gas cell. The heat flow rate is described by:

$$\dot{Q} = \alpha\,A \cdot \Delta T \ . \tag{3.25}$$

where ΔT is the temperature difference between the ports and αA is the heat transfer coefficient. For the heat transfer coefficient, any type of correlation may be used. The heat transfer coefficients used for the adsorption chiller in this thesis are further specified in Section 5.1.

3.3.2 Mass transfer

The mass transfer model connects two cells via their fluid ports and determines the mass flow between the cells. For convective flows, the mass flow is related to the pressure drop Δp by a correlation dependent on fluid properties and geometry. For convective flows, the mass transfer model reads:

$$\dot{m} = \beta_p \Delta p \ . \tag{3.26}$$

In this thesis, the convective mass transfer model is used to describe the inter-particle mass transfer resistance (cf. Section 2.2.2). The used pressure drop correlation is described in Section 5.1.

Additionally, a mass transfer model is implemented which uses the loading difference between actual loading w_{ad} and equilibrium loading w_{eq} as driving potential.

$$\dot{m} = \beta_w \left(w_{\text{ad}} - w_{\text{eq}} \right) \ . \tag{3.27}$$

This type of mass transfer model is used to describe the intra-particle mass transfer resistance. In this thesis, the intra-particle mass transfer resistance is determined by the Linear Driving Force model, first introduced by Glueckauf (1955), which is described in detail in Section 5.1. The mass transfer coefficient is denoted β_w.

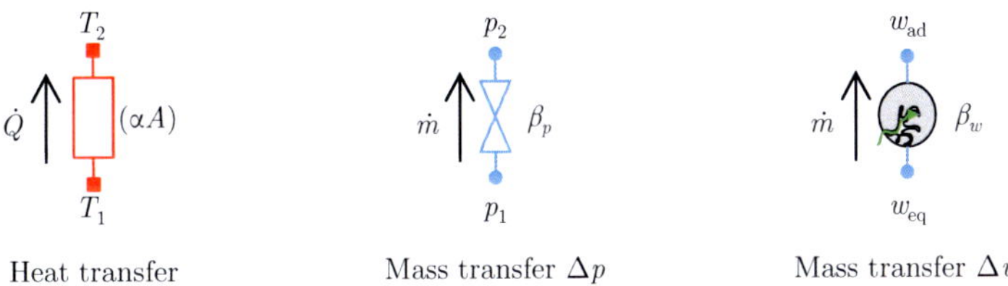

Figure 3.3: Scheme of heat and mass transfer functions. A flow variable is calculated ($\dot{Q}$, $\dot{m}$) dependent on the potential variables (T, p, and w) and transfer coefficients (αA, β_p, and β_w).

3.4 Illustration examples

The presented Adsorption Energy Systems Library allows to modularly build adsorption-based devices. To demonstrate library flexibility and achieved model accuracy, three examples are presented: a desiccant unit, an adsorption thermal storage unit, and an adsorption chiller. For the design context of an adsorption chiller in this thesis, the library allows to change working pairs, component designs and cycle designs in a fast and convenient manner.

3.4.1 Desiccant unit

An adsorption dryer or desiccant unit is used to simultaneously reduce air humidity and preheat the air. During adsorption, cold humid air enters the dry adsorber. Until equilibrium state is reached water is adsorbed and the adsorption enthalpy is released. For regeneration of the unit by desorption, hot dry air flows through the adsorber. The air leaves the adsorber with a higher water loading and a decreased temperature.

The adsorber model created within the Adsorption Energy Systems Library is validated using experimental data by Pesaran and Mills (1987a). The used adsorber has a cylindrical shape (diameter $d_i = 0.13\,\mathrm{m}$, length $l = 0.05\,\mathrm{m}$) and is filled with silica gel (Equilibrium data of silica gel is used from Pesaran and Mills (1987b)).

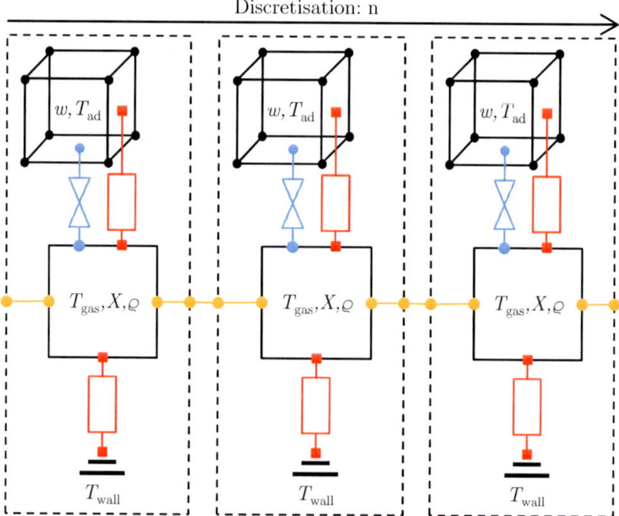

Figure 3.4: Scheme of desiccant unit consisting of an adsorbent cell, a gas cell, and a heat capacity. The cells are connected by heat and mass transfer resistances with coefficients from Hougen and Marshall (1947). The desiccant unit is discretised in flow direction.

The driving potential used for mass transfer is pressure difference Δp. Coefficients for mass and heat transfer are modelled by correlations dependent on the Reynolds number based on Hougen and Marshall (1947). Since the model determines the outlet temperature and mass fraction from the inlet values, geometries and input parameters, the model is predictive.

Figures 3.5a and 3.5b show temperature and water mass fraction during an adsorption process at adsorber inlet and outlet.

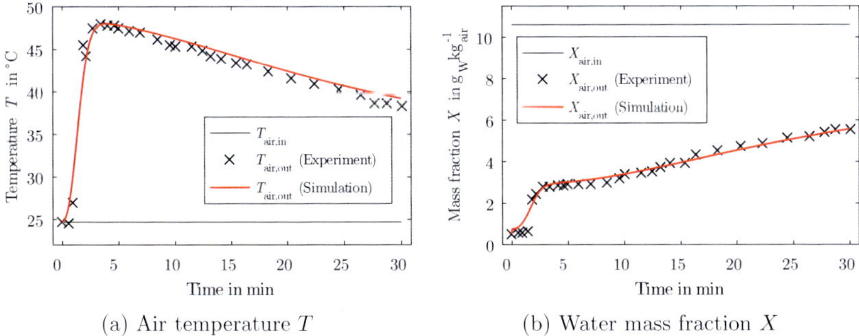

(a) Air temperature T

(b) Water mass fraction X

Figure 3.5: Air temperature T_{air} and water mass fraction X_{air} at inlet and outlet during adsorption. The experimental data shown are reported by Pesaran and Mills (1987a), (experiment 7).

Validation shows good agreement in both, temperature T_{air} and mass fraction X_{air}. The input conditions are changed suddenly at $t = 0\,\mathrm{s}$, resulting in a step function. The output is reacting by an increase in outlet temperature due to emitted adsorption heat. Around 5 min after start, temperature peaks at 46.9 °C and afterwards declines only slightly. By reaching a temperature of 45.4 °C, simulation corresponds well to the measured peak after around 5 min. Water mass fraction between inlet and outlet is reduced by about 10 $g_{water}kg_{air}^{-1}$ with nearly dry air at the outlet. Measured and simulated water loading fit almost perfectly. Pesaran and Mills (1987a) have experimentally varied input parameters, adsorber length, and particle size of adsorbent grains in their experiments. For all six tested input variations, simulation and experiment agreed equally well (not shown).

For systems including a desiccant unit, the Adsorption Energy Systems Library is used within the following publications: Bau et al. (2015c,d, 2016b,c); Erdogan et al. (2017).

3.4.2 Thermal adsorption storage unit

Thermal energy, stored in an adsorption storage unit, can be divided into a sensible and a latent part. Sensible heat increases with system temperature, while, at the same time, latent heat is stored by desorption of water. The desorbed water leaves the adsorber as vapour and is condensed afterwards in the condenser. During condensation, low grade heat is emitted at medium or low temperature level, that can be either used or released to the ambient. When discharging the storage unit, water is evaporated using low grade heat. The vaporised water flows to the adsorber where it is adsorbed and heat at process temperature is released.

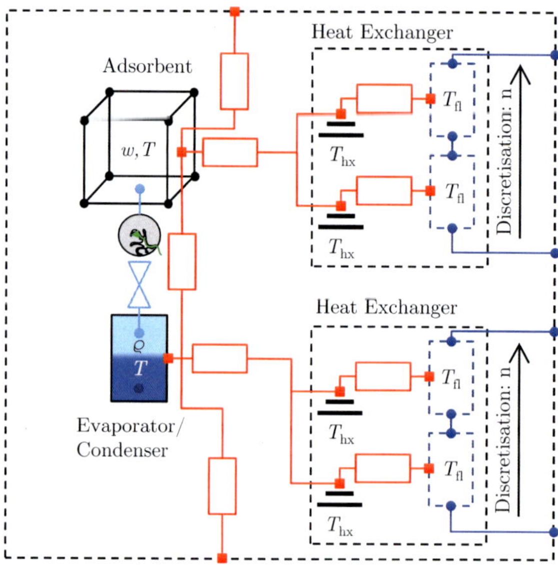

Figure 3.6: Scheme of thermal adsorption storage unit consisting of an adsorbent, two heat exchangers (right), mass resistances, heat resistances, and an evaporator/condenser model. The heat exchangers are discretised in flow direction.

The model of the adsorption storage unit is validated using experimental data from a storage prototype at the Institute of Technical Thermodynamics at RWTH Aachen University (Schreiber et al., 2014). Figure 3.6 illustrates the model. Since the temperature difference between adsorber and evaporator is high and there is no insulation between these components, they are thermally connected in the model. For further details regarding the experimental setup see Binkert et al. (2013); Schreiber (2017).

The working pair used in the experiments is zeolite 13X / water. Equilibrium data for this pair can be found in Núñez (2001). Heat and mass transfer coefficients were determined by calibration with experimental data. Figure 3.7a shows the adsorber heat flow rate during desorption, Figure 3.7b during adsorption. Desorption temperature increases to almost 200 °C, with a condensation temperature of 90 °C. After around 3 hours, the operation mode is switched from desorption to adsorption (loading to unloading). Adsorption takes about 30 min with adsorption temperature falling from 200 °C to 120 °C. Vaporisation is kept at 60 °C during adsorption. Although time varies by factor 6 and heat flow rate even by factor 10, the comparison between experimental data and simulations shows good agreement for both desorption and adsorption. During desorption, the peak in heat flow rate at 15 min is slightly overestimated by the simulation; for adsorption, it can be observed that the trend fits well. Only slight differences can be observed between the experiment and the simulation at the end of adsorption phase.

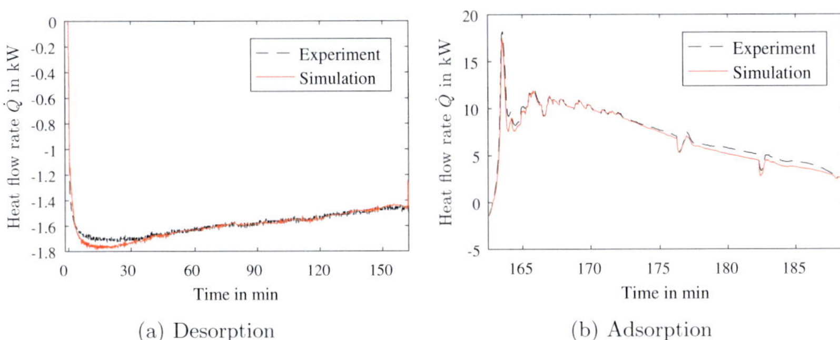

(a) Desorption

(b) Adsorption

Figure 3.7: Measured and simulated heat flow rate $\dot{Q}$ of adsorber during desorption and adsorption (Schreiber et al., 2014).

Concerning adsorption thermal storage systems, the Adsorption Energy Systems Library is used within the following publications: Schreiber et al. (2015, 2016a,b).

3.4.3 Adsorption chiller

The adsorption chiller consists of an adsorber, an evaporator, and a condenser in the simple one-bed configuration. The components are separated by valves, allowing to separately disconnect the adsorber from the evaporator and condenser (Figure 3.8).

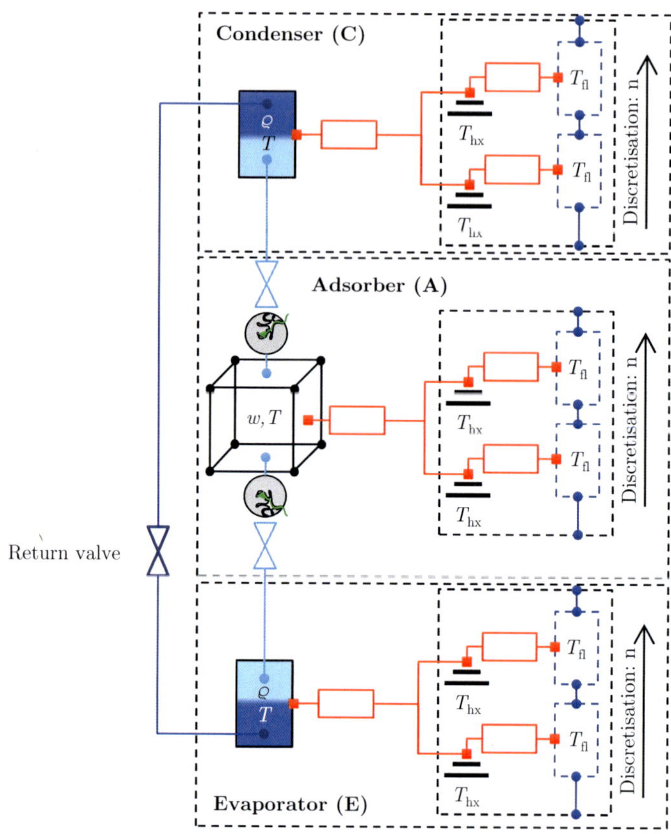

Figure 3.8: Scheme of adsorption chiller consisting of an adsorbent, evaporator, condenser, mass resistances, and heat resistances. The heat exchangers are 1-d discretised in flow directions. Condenser and evaporator are connected by a return valve.

This configuration is experimentally investigated for both silica gel and zeolite 13X as adsorbent. In this short example, the validation for silica gel 123 / water is shown. A more in-depth validation can be found in Lanzerath et al. (2015), discussing the effects of varying input parameters and changed adsorbent materials.

Equilibrium data for silica gel 123 / water for simulation are used from Schawe

(1999). Coefficients for heat and mass transfer are used as fitting parameters. They are determined by using the experimental data of a complete cycle (adsorption and desorption) and minimizing the coefficient of variation (CV) which evaluates the deviation between measured and simulated heat flow rates. The fitted heat and mass transfer parameters were kept constant for all variations of experimental settings and proved to be robust: $CV_{\text{calibration}} = 13.6\,\%$ and $CV_{\text{validation}} = 12.6\,\%$ to $27.4\,\%$ (Lanzerath et al., 2015). The determined parameters are also used in the following chapters of this thesis.

Figure 3.9a shows the adsorber heat flow rate and Figure 3.9b the heat flow rates of evaporator and condenser. Although only lumped models are used for adsorber, evaporator and condenser, the system dynamics, as well as steady state conditions, are predicted well.

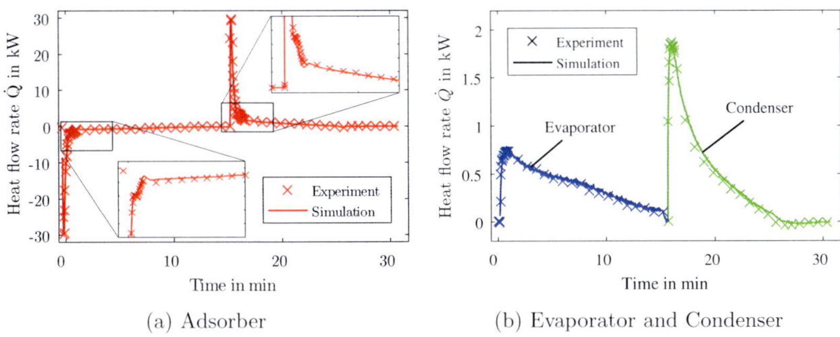

(a) Adsorber

(b) Evaporator and Condenser

Figure 3.9: Heat flow rates of experiment and simulation at the three heat exchangers: adsorber, evaporator, and condenser. The figure shows the calibration cycle as reported by Lanzerath et al. (2015).

Concerning adsorption chillers, the Adsorption Energy Systems Library is used within the following publications: Bau et al. (2015a,b); Lanzerath et al. (2015); Bau et al. (2016a, 2017a,b,c).

In addition, the library is used within the following works: prediction of heat and mass transfer coefficients with the help of infra-red large-temperature jump experiments (Graf et al., 2015, 2017b); prediction of SCP and COP in combination with gravimetric large-temperature jump experiments (Graf et al., 2016); evaluation of a hybrid energy system consisting of a CO_2 compression chiller and an adsorption chiller (Gibelhaus et al., 2017); and analysing the optimal adsorbent for cooling applications (Graf et al., 2017a).

Part II

Design of adsorption energy systems

Chapter 4

Framework to optimise design ensuring optimal control

In Chapter 2, the main challenges in designing adsorption chillers have been presented: the influence of input conditions, the dependency on control, the trade-off between efficiency and power density, and the interdependency between the design levels. In Chapter 3, a dynamic-model library in the modelling language Modelica is presented which allows to quickly model and simulate new adsorption energy systems, e. g. adsorption chiller with different cycle designs. In this chapter, a model-based framework to thoroughly assess, compare, and optimise adsorption chiller designs is presented. The framework explicitly addresses all interdependencies and difficulties presented in Chapter 2 and allows to evaluate design choices in the objective space of an adsorption chiller.

The framework can be used to answer the two following types of questions:

- How do design choices compare under optimal control? (Problem 1)
- What is the optimal design and control? (Problem 2)

Depending on the type of question, the optimisation parameters have to be varied. Figure 4.1 illustrates the 3 steps of the framework: (1) setting up a dynamic model, (2) formulating and solving a multi-objective optimisation problem, and (3) analysing and interpreting the obtained Pareto frontiers to compare and optimise adsorption chiller designs.

Step 1: The dynamic model is the basis to set up the optimisation problem. The model has to fulfil two main requirements: First, the model has to be valid in the range of the investigated parameters. This can be ensured by calibration and validation with experimental data. Second, the model has to be computationally sufficiently fast for optimisation, but detailed enough to incorporate all design effects to be optimised.

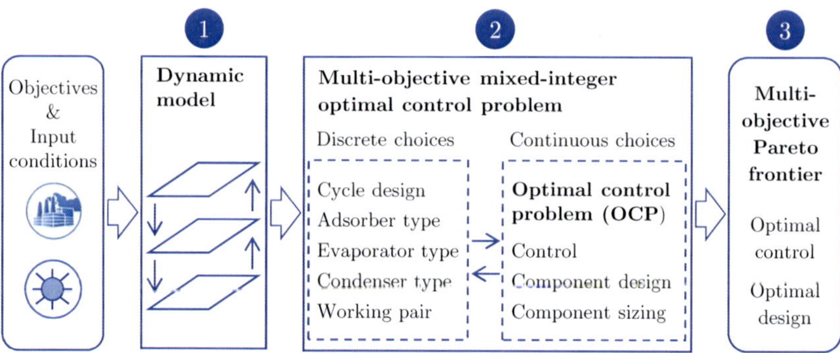

Figure 4.1: Steps of the proposed framework for rigorous model-based assessment and optimisation of adsorption chillers. (1) A dynamic model is needed which fully describes the adsorption chiller's behaviour and captures the effects of the investigated design choices. (2) A multi-objective mixed-integer optimal control problem is set up and solved including control and design choices. (3) Solving the optimisation problem gives a Pareto frontier for all objectives allowing to compare design. Additionally, the values for optimal control and optimal design are obtained.

Step 2: The task of finding the optimal control and design can be translated into a multi-objective optimisation problem. This optimisation problem varies, depending on the initial type of question.

If designs are to be assessed and compared (Problem 1), the optimisation problem only includes control parameters as optimisation variables. Optimisation of control ensures that the intrinsic characteristics of each design are investigated and not the effect of poor control. In this case, the optimisation problem belongs to the class of optimal control problems (OCPs).

If design and control are to be optimised simultaneously (Problem 2), the problem class depends on the design variables which are included. The optimisation problem still belongs to the class of OCPs, if only continuous design parameters are included. If also discrete design choices are included, the problem belongs to the class of mixed-integer optimal control problems. Possible continuous design variables are geometry parameters like grain size or heat exchanger length. Discrete design choices are the working pair and the cycle design.

In this thesis, OCPs are solved by an efficient multiple shooting algorithm (Leineweber et al., 2003). Mixed-integer OCPs are not solved algorithmically, but rather multiple discrete design choices are compared. In this thesis, 2 working pairs and 5 cycle designs are considered. For each discrete design choice, the optimal control sub-problem of continuous design choices and control is solved.

Step 3: In a last step, the optimisation problem is solved and the resulting Pareto frontiers are analysed. Based on the designer's preferences on efficiency and power density, an optimal design can be chosen from the obtained solutions.

In general, it is desirable to solve the multi-objective mixed-integer optimal control problem in a single optimisation including all discrete and continuous design choices. This single optimisation would lead to the overall Pareto frontier for given objectives and input conditions. This overall Pareto frontier shows the best achievable performance and can be interpreted as an envelope; all Pareto frontiers obtained by solving sub-problems (not all design choices are included) must lie within this envelope.

Since capturing all design choices within one single problem formulation is a difficult and complex task, in this thesis, complexity of the optimisation problem grows successively in the Chapters 5 - 8. Additionally, effects of input conditions and objectives on optimal design and control are investigated using the framework.

In Chapter 5, the framework is used to optimise the adsorber-bed design of a finned-tube heat exchanger with loose silica gel grains. For optimisation, grain size and fin number are considered. In this chapter, each step of the framework is described in detail. Additionally, the loss in adsorption potential is introduced as a way to analyse designs.

In Chapter 6, the optimisation problem is enhanced by including component sizing: the ratios between adsorber, evaporator, and condenser are added to the optimisation problem. Additionally, the effect of objectives on the optimal design is investigated: a specific cooling power only considering the adsorbent mass (SCP_{ads}) is compared to a specific cooling power considering the mass of all components (SCP_{all}).

In Chapter 7, the effect of working pair on optimal design and control is investigated. For this purpose, the optimisation problem is enhanced by a second working pair: AQSOA Z02 / water. In this chapter, also, the evaporator input temperature is varied and the effects are analysed.

Finally, in the last chapter of Part II, 5 cycle designs for two-bed chillers are compared. To do so, first a thermodynamic model (cf. Section 2.4.1) is used to determine the maximum efficiency of each cycle design. Afterwards, a multi-objective optimal control problem is solved for each cycle design to obtain the Pareto frontiers.

Chapter 5

Simultaneous optimisation of adsorber-bed design and control

In this chapter, the proposed framework is exemplified by optimising the adsorber-bed design of a heat exchanger with loose silica gel grains proposed by Lanzerath et al. (2015). One-bed simple cycle and ideal evaporator and condenser characteristics are used for the other discrete choices (cf. Figure 4.1. The three steps of the framework are described in detail: (1) setting up the dynamic model, (2) formulating and solving the multi-objective optimal control problem, and (3) analysing and interpreting the obtained Pareto frontiers.

In Section 5.1, a dynamic model of the adsorption chiller is derived. The starting point is a model which is experimentally calibrated and validated by Lanzerath et al. (2015). This initial model is enhanced to capture the effect of varying grain size and fin number.

In Section 5.2, a multi-objective optimal control problem (OCP) is presented. This multi-objective OCP allows to assess designs at optimal control and also to simultaneously optimise control and design.

Contents of this chapter have been published in:

Bau, U., Hoseinpoori, P., Graf, S., Schreiber, H., Lanzerath, F., Kirches, C., and Bardow, A. (2017). Dynamic optimisation of adsorber-bed designs ensuring optimal control. *Applied Thermal Engineering*, 125:1565-1576.

In Section 5.3, the optimal control of the initial design is presented. This design is further analysed by introducing the loss of adsorption potential, which is a way to quantify the contributions of all heat and mass transfer resistances. Afterwards, the effect of a varying grain size and fin number on performance is successively analysed. This is done by comparing specific choices of grain size and fin number while ensuring optimal control. Thereby, only the effect of the design parameters and not the effect of poor control is captured. In a last step, the simultaneous optimisation of control and design is conducted which also includes grain size and fin number as optimisation parameters.

Figure 5.1 shows all discrete choices and the considered optimisation parameters of this chapter.

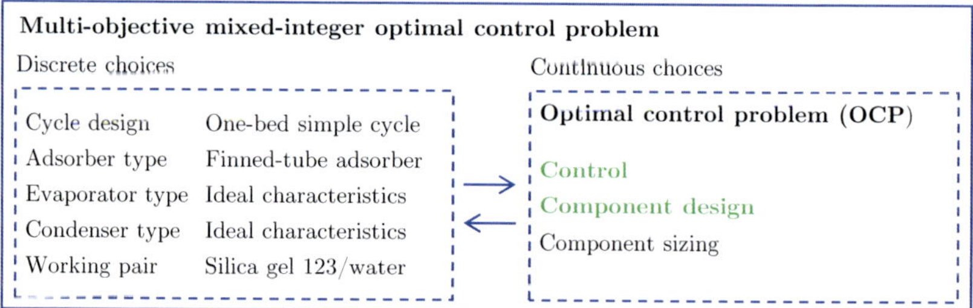

Figure 5.1: Discrete choices and considered optimisation parameters (green) for initial case study: optimisation of adsorber-bed design with ideal evaporator and condenser and one-bed simple cycle.

5.1 Dynamic model and model calibration

In this section, we derive the dynamic model used to investigate a finned-tube adsorber-bed design filled with silica gel grains. The initial design is the adsorber bed experimentally studied by my co-workers and me (Lanzerath et al., 2015). For the model of this initial design, the general structure is presented followed by the heat and mass transfer resistances taken into account, the chosen model boundaries, and the determined model parameters (Lanzerath et al., 2015). Afterwards, the specific model equations are introduced which are used to capture the design choices, grain size and fin number, on system's performance.

The model of the adsorption chiller (Figure 5.2) mainly consists of models for the adsorbent material and the adsorber heat exchanger. These models are connected by a heat transfer resistance model (α_{ad}). The adsorber-bed heat exchanger is also coupled to a fluid circuit by a heat transfer resistance model (α_{hx}). The fluid circuit input conditions are an input temperature $T_{\text{fl,in}}$ and a constant volume flow rate $\dot{V}_{\text{fl}}$. The adsorbent material is connected to the evaporator and condenser by mass transfer resistance models. Both inter-particle (β_{Darcy}) and intra-particle (β_{LDF}) mass transfer are considered in the model (Figure 5.4).

To consider only the intrinsic characteristics of the adsorber bed, all other components are represented by ideal characteristics: In case of evaporator and condenser, the saturation pressures ($p_{\text{E}} = p_{\text{s}}(T_{\text{E}})$ and $p_{\text{C}} = p_{\text{s}}(T_{\text{C}})$) are directly linked to the adsorber bed via mass transfer resistance models (β_{Darcy} and β_{LDF}). Thus, all negative effects are excluded which impact overall system's performance due to heat transfer limitations in evaporator and condenser.

As a difference to the model used by Lanzerath et al. (2015), the discretisation in the heat exchanger is removed. Lanzerath et al. (2015) discretised the adsorber heat exchanger with 40 elements to model the plug flow in the heat exchanger as accurately as possible which was found necessary for calibration. For optimisation, as conducted in this thesis, a low number of states is beneficial regarding computational time. Therefore, the heat exchanger discretisation is removed and a linear temperature distribution in the heat exchanger is assumed to compensate for the low discretisation. The resulting error in efficiency (COP) and power density (SCP) is less than 1 % and, therefore, the error is regarded as acceptable. For more information on the used discretisation and temperature distribution see Appendix B.

The model is built in Modelica with the LTT Adsorption Energy Systems Library presented in Chapter 3. The resulting model is solved for four differential states: adsorbent loading w_{ad}, adsorbent temperature T_{ad}, heat exchanger temperature T_{hx}, and fluid temperature T_{fl}. The main model equations are summarised in Table 5.1. For additional information on the derivation of the model equations, the reader is referred to Chapter 3.

To consider the changing model equations and inlet conditions between adsorption and desorption, the model consists of two stages, representing the adsorption phase and the desorption phase (cf. Section 2.1.2). Isosteric cooling is included in the adsorption phase, isosteric heating in the desorption phase. For the adsorption phase, the inlet temperature is set to medium temperature level ($T_{\text{fl,in}} = T_{\text{mid}}$), for desorption, the inlet temperature is set to high temperature level ($T_{\text{fl,in}} = T_{\text{high}}$). The pressures at evaporator and condenser (p_{E} and p_{C}) are constant during the entire adsorption cycle. The division into two model stages (adsorption and desorption) is also relevant for optimisation, which is further explained in Section 5.2.

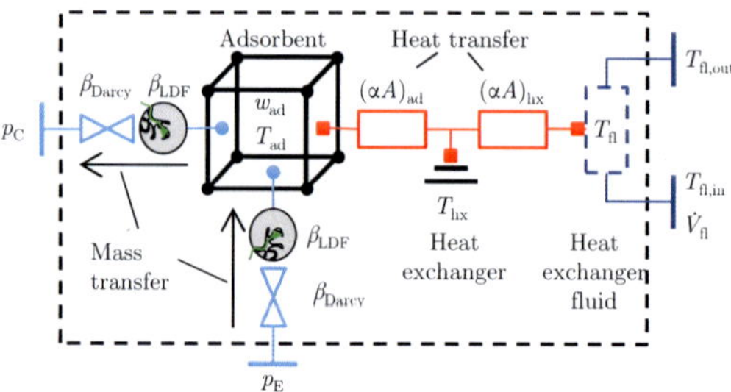

Figure 5.2: Structure of the dynamic adsorber-bed model: adsorbent and heat exchanger, connected by models for heat and mass transfer.

The heat and mass transfer coefficients and the equilibrium data have to be determined experimentally, taken from the literature, or derived from more sophisticated models. Geometrical and material data (e. g. heat exchanger surface A_{ad}, adsorbent mass m_{sor}, ...) can be directly calculated from the adsorber-bed design. Additionally, two characteristic ratios are introduced to capture the effects of design choices: the ratio of heat exchanger surface area to adsorbent mass $A_{\text{ad}}/m_{\text{sor}}$ and the heat capacity ratio of heat exchanger to adsorbent $C_{\text{hx}}/C_{\text{sor}} = (m_{\text{hx}}c_{\text{hx}})/(m_{\text{sor}}c_{\text{sor}})$. These ratios are normalised by the adsorbent mass and thus, allow for a comparison of differently sized systems.

For the initial adsorber-bed design (Figure 5.3a), Lanzerath et al. (2015) determined the heat and mass transfer coefficients by a full-scale experiment of a one-bed adsorption chiller. The parameters are listed in Table 5.2. For the used working pair, silica gel 123 / water, the equilibrium data is described by the model of Dubinin (1967) with a characteristic curve measured by Schawe (1999). Figure 5.3b shows a comparison between measured and simulated cooling power to illustrate the high model accuracy achieved by Lanzerath et al. (2015) (also cf. Section 3.4.3).

To allow for design optimisation, the dynamic model has to capture the effects of the design choices, which are to be investigated. In this chapter, the design choices are the grain size and the fin number. The effect of grain size on the performance is captured by the inter-particle β_{Darcy} and intra-particle β_{LDF} mass transfer resistances. The effect of fin number is captured by the mass of metal m_{hx}, the mass of adsorbent m_{sor}, and the heat transfer coefficient $(\alpha A)_{\text{ad}}$. In the following, specific model equations are derived to calculate the influence of grain size and fin number on system's performance.

Table 5.1: Overview of model equations to describe the adsorber bed. The mass transfer coefficients for inter-particle β_{Darcy} and intra-particle β_{LDF} mass transfer are described in detail in the text.

Mass balance adsorbent	$\frac{\mathrm{d}m_{\text{ad}}}{dt} = m_{\text{sor}}\frac{\mathrm{d}w_{\text{ad}}}{dt}$	$=$	$\dot{m}_{\text{E}} - \dot{m}_{\text{C}}$	
Energy balance adsorbent	$u_{\text{ad}}(T_{\text{ad}}, w_{\text{ad}})\frac{\mathrm{d}m_{\text{sor}}}{dt} + (m_{\text{ad}}c_{\text{ad}} + m_{\text{sor}}c_{\text{sor}})\frac{\mathrm{d}T_{\text{ad}}}{dt}$	$=$	$\dot{m}_{\text{E}}h_{\text{v}}(T_{\text{E}}, p_{\text{E}}) - \dot{m}_{\text{C}}h_{\text{v}}(T_{\text{ad}}, p_{\text{ad}}) + \dot{Q}_{\text{hx-ad}}$	
Energy balance heat exchanger	$m_{\text{hx}}c_{\text{hx}}\frac{\mathrm{d}T_{\text{hx}}}{dt}$	$=$	$\dot{Q}_{\text{fl-hx}} - \dot{Q}_{\text{hx-ad}}$	
Mass balance heat exchanger fluid	0	$=$	$\dot{m}_{\text{fl,in}} - \dot{m}_{\text{fl,out}}$	
Energy balance heat exchanger fluid	$m_{\text{fl}}c_{\text{fl}}(T_{\text{fl}})\frac{\mathrm{d}T_{\text{fl}}}{dt}$	$=$	$\dot{m}_{\text{fl,in}}c_{\text{fl}}(T_{\text{fl}})(T_{\text{fl,in}} - T_{\text{fl,out}}) - \dot{Q}_{\text{fl-hx}}$	
Heat transfer	$\dot{Q}_{\text{fl-hx}}$	$=$	$(\alpha\text{A})_{\text{hx}}\,(\bar{T}_{\text{fl}} - T_{\text{hx}})$	
	$\dot{Q}_{\text{hx-ad}}$	$=$	$(\alpha\text{A})_{\text{ad}}\,(T_{\text{hx}} - T_{\text{ad}})$	
Mass transfer: (E) → (A)	$\dot{m}_{\text{E}}$	$=$	$\beta_{\text{Darcy}}[p_{\text{E}} - p_{\text{E}}^{*}]$,	for $p_{\text{E}} > p_{\text{ad}}$
		$=$	$\beta_{\text{LDF}}m_{\text{sor}}\left[w_{\text{eq}}(p_{\text{E}}^{*}, T_{\text{ad}}) - w_{\text{ad}}\right]$,	
	$\dot{m}_{\text{E}}$	$=$	0,	for $p_{\text{E}} \leq p_{\text{ad}}$
	p_{E}^{*}	$=$	$\frac{1}{2}(p_{\text{E}} + p_{\text{ad}})$	
Mass transfer: (A) → (C)	$\dot{m}_{\text{C}}$	$=$	$\beta_{\text{LDF}}m_{\text{sor}}\left[w_{\text{ad}} - w_{\text{eq}}(p_{\text{C}}^{*}, T_{\text{ad}})\right]$,	for $p_{\text{C}} < p_{\text{ad}}$
		$=$	$\beta_{\text{Darcy}}[p_{\text{C}}^{*} - p_{\text{C}}]$,	
	$\dot{m}_{\text{C}}$	$=$	0,	for $p_{\text{C}} \geq p_{\text{ad}}$
	p_{C}^{*}	$=$	$\frac{1}{2}(p_{\text{C}} + p_{\text{ad}})$	
Coefficient of performance (COP)	COP	$=$	$\frac{\int_0^{t_{\text{ads}}} \dot{m}_{\text{E}}\left(h_{\text{s}}^{\text{v}}(T_{\text{E}}) - \left(h_{\text{s}}^{\text{l}}(T_{\text{C}}) - h_{\text{s}}^{\text{l}}(T_{\text{E}})\right)\right)\mathrm{d}t}{\int_{t_{\text{ads}}}^{t_{\text{des}}} \dot{m}_{\text{fl}}c_{\text{fl}}(T_{\text{fl,in}} - T_{\text{fl,out}})\,\mathrm{d}t}$	
Specific cooling power (SCP_{ads})	SCP_{ads}	$=$	$\frac{\int_0^{t_{\text{ads}}} \dot{m}_{\text{E}}\left(h_{\text{s}}^{\text{v}}(T_{\text{E}}) - \left(h_{\text{s}}^{\text{l}}(T_{\text{C}}) - h_{\text{s}}^{l}(T_{\text{E}})\right)\right)\mathrm{d}t}{m_{\text{sor}}}$	
Inlet conditions	$T_{\text{fl,in}}$	$=$	T_{mid}	for $0 \leq t < t_{\text{ads}}$
	$T_{\text{fl,in}}$	$=$	T_{high}	for $t_{\text{ads}} \leq t < t_{\text{des}}$
	p_{E}	$=$	$p_{\text{s}}(T_{\text{low}})$	
	p_{C}	$=$	$p_{\text{s}}(T_{\text{mid}})$	

Table 5.2: Initial design parameters determined by Lanzerath et al. (2015).

Parameter	Value	
$A_{\mathrm{ad}}/m_{\mathrm{sor}}$	9.7×10^{-1}	$\mathrm{m^2\,kg^{-1}}$
β_{LDF}	1.33×10^{-4}	$\mathrm{s^{-1}}$
β_{Darcy}	∞	$\mathrm{kg\,s^{-1}\,Pa^{-1}}$
$C_{\mathrm{hx}}/C_{\mathrm{sor}}$	1.77	
α_{ad}	1.53×10^{2}	$\mathrm{W\,m^{-2}\,K^{-1}}$
A_{ad}	1.82	$\mathrm{m^2}$
$(\alpha A)_{\mathrm{hx}}$	2.26×10^{3}	$\mathrm{W\,K^{-1}}$
n_{fins}	14	
d_{p}	9×10^{-3}	m

(a) Adsorber bed

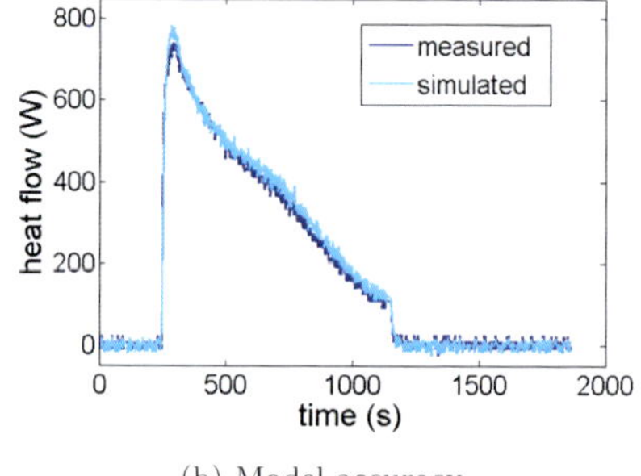

(b) Model accuracy

Figure 5.3: Adsorber-bed design investigated by Lanzerath et al. (2015). (a) Experimentally investigated adsorber bed and heat exchanger tubes. (b) Comparison between measured and simulated cooling power for the calibrated adsorber-bed model.

The inter-particle mass transfer resistance describes the pressure drop that the adsorbate vapour experiences while penetrating the adsorbent layer and reaching the particle's surface. The main driving force for this mass transfer is the pressure gradient within the packed bed of adsorbent grains. The phenomenon can be described by Darcy's law (Bejan, 2013). Using Darcy's law, the vapour velocity v can be calculated as:

$$v = -\frac{K}{\mu}\nabla p \tag{5.1}$$

where μ describes the viscosity of the adsorbate vapour and K the permeability of the adsorbent bed. The permeability is a function of particle diameter d_p and porosity of the bed ϵ_b. According to Ben Amar et al. (1996), for spherical grains, the permeability can be calculated as:

$$K = \frac{d_\mathrm{p}^2 \cdot \epsilon_\mathrm{b}^3}{150(1-\epsilon_\mathrm{b})^2} \,. \tag{5.2}$$

Rewriting Equation (5.1) for a lumped model considering mass transport within the adsorbent layer in radial direction leads to:

$$\dot{m}_\mathrm{Darcy} = v A_\mathrm{c} \varrho_\mathrm{v} = -\underbrace{\frac{K \cdot A_\mathrm{c} \cdot \varrho_\mathrm{v}}{\mu \cdot \delta}}_{\beta_\mathrm{Darcy}} (p_\mathrm{E(C)} - p^*) \,, \tag{5.3}$$

where A_c is the mean cross sectional area of the adsorbent bed the adsorbate vapour has to pass and ϱ_v is the density of the adsorbate vapour. The pressure p^* represents the mean pressure at the outer surface of the particles. It is assumed that the vapour has to pass half of the adsorbent mass on average. For a circular tube, this assumption can be translated into the length δ which the vapour has to travel (Figure 5.4b).

The second resistance for the adsorbate vapour is the intra-particle resistance occurring inside the particles. To account for all resistances within the grain, the linear driving force (LDF) model is used (Bathen, 2001), with the loading gradient as driving force:

$$\frac{\partial w}{\partial t} = \beta_\mathrm{LDF}(w_\mathrm{eq}(p^*, T_\mathrm{ad}) - w_\mathrm{ad}) \,, \tag{5.4}$$

where w_eq is the equilibrium uptake at (p^*, T_ad). According to Glueckauf (1955), for a perfect sphere, β_LDF can be calculated as:

$$\beta_\mathrm{LDF} = \frac{60D}{d_\mathrm{p}^2} \,, \tag{5.5}$$

where D is the effective diffusion coefficient.

The relative size of the pressure at the particle surface p^* compared to the evaporator pressure $p_\mathrm{E(C)}$ and adsorbent equilibrium pressure p_ad indicates whether inter-particle or intra-particle mass transfer resistance is dominating. This effect is mainly governed by the grain size as design variable.

The fin number directly affects the mass of metal m_hx, the mass of adsorbent m_sor, and the heat transfer area between heat exchanger and adsorbent material A_ad. To capture these effects, the necessary correlations are derived from geometrical information based on the initial design (cf. Figure 5.3a and 5.4).

The mass of heat exchanger metal is split into the inner tube, which is independent of the fin number and the fins. Thus, the mass of heat exchanger metal can be

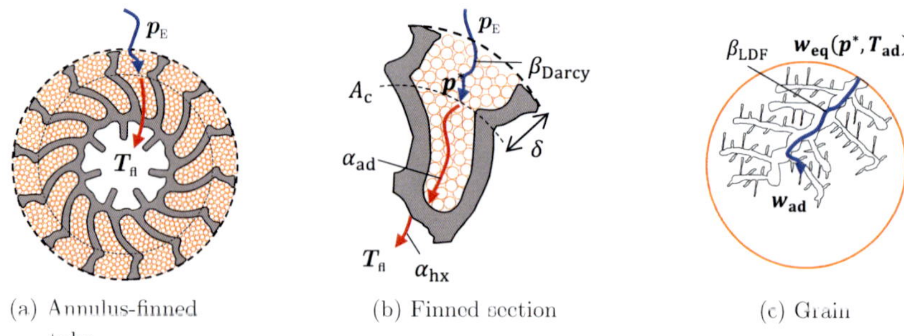

(a) Annulus-finned tube (b) Finned section (c) Grain

Figure 5.4: Schematics of (a) annulus-finned tube, (b) finned section, and (c) grain. The indicated heat and mass transfer coefficients correspond to the model shown in Figure 5.2 and the equations in Table 5.1. The indicated flow directions represent the case of adsorption.

calculated as

$$m_{\text{hx}} = (V_{\text{tube}} + n_{\text{fins}} V_{\text{fin}})\, \varrho_{\text{hx}} \,, \tag{5.6}$$

where V_{tube} is the volume of the plain tube without fins, V_{fin} is the volume of a fin, and ϱ_{hx} is the density of the heat exchanger material.

The mass of adsorbent is calculated by multiplying the volume not occupied by fins V_{sor} with the bulk density of the used adsorbent material $\varrho_{\text{sor,bulk}}$:

$$m_{\text{sor}} = V_{\text{sor}} \varrho_{\text{sor,bulk}} = (V_{\text{outer,total}} - n_{\text{fins}} V_{\text{fin}})\, \varrho_{\text{sor,bulk}} \,, \tag{5.7}$$

where $V_{\text{outer,total}}$ is the maximum volume on the outside of the tube which can be filled with adsorbent material when assuming no fins and a constant outer radius. The bulk density of the adsorbent is calculated from the initial design and is assumed to be constant.

Finally, the heat transfer area between finned tube and adsorbent A_{ad} is calculated. To do so, the surface area of the plain tube $A_{\text{outer,tube}}$ is taken as base area. The specific surface area A_{fin} of each fin is added. Additionally, the area lost due to mounting the fin on the tube is subtracted. The lost area can be expressed by the thickness of the fin t_{fin} multiplied by the length of the heat exchanger l_{hx}. The resulting equation reads:

$$A_{\text{ad}} = A_{\text{outer,tube}} + n_{\text{fins}} \left(A_{\text{fin}} - t_{\text{fin}} l_{\text{hx}}\right) \,. \tag{5.8}$$

All additionally used geometric and mass transfer parameters are listed in Table 5.3.

Table 5.3: Additional geometric and mass transfer parameters used to describe effects of grain size and fin number on system's performance. Geometric parameters are normalised by heat exchanger length l_{hx} to allow for a comparison of differently sized systems.

Parameter	Value
$V_{\mathrm{tube}}/l_{\mathrm{hx}}$	8.36×10^{-5} $\mathrm{m^3\,m^{-1}}$
$V_{\mathrm{fin}}/l_{\mathrm{hx}}$	1.48×10^{-5} $\mathrm{m^3\,m^{-1}}$
$V_{\mathrm{outer,total}}/l_{\mathrm{hx}}$	7.73×10^{-4} $\mathrm{m^3\,m^{-1}}$
ϱ_{hx}	2.70×10^{3} $\mathrm{kg\,m^{-3}}$
$\varrho_{\mathrm{sor,bulk}}$	7.61×10^{2} $\mathrm{kg\,m^{-3}}$
$A_{\mathrm{outer,tube}}/l_{\mathrm{hx}}$	1.95×10^{-2} $\mathrm{m^2\,m^{-1}}$
$A_{\mathrm{fin}}/l_{\mathrm{hx}}$	2.51×10^{-2} $\mathrm{m^2\,m^{-1}}$
$A_{\mathrm{c}}/l_{\mathrm{hx}}$	8.64×10^{-2} $\mathrm{m^2\,m^{-1}}$
t_{fin}	1.13×10^{-3} m
D	1.8×10^{-10} $\mathrm{m^2\,s^{-1}}$
ϵ_{b}	0.32

5.2 Multi-objective DAE-constrained optimal control problem

The dynamic model described in Section 5.1 consists of a set of differential and algebraic equations (DAEs). In order to find the optimal control and design, a multi-objective optimal control problem (OCP) is set up and solved containing the DAE-system as a constraint. In this section, first, the arising optimal control problem is derived stepwise and the resulting problem class is stated. Afterwards, the direct multiple shooting method is presented, first introduced by Bock and Plitt (1984), as a state-of-the-art direct and all-at-once approach to non-linear optimal control.

5.2.1 Optimal control problem

Modelling of a dynamic system, e. g. the adsorption chiller described in Section 5.1, generally leads to a set of differential and algebraic equations (DAEs). In the semi-explicit form, this set of DAEs can be written as

$$\dot{x}(t) = f(x(t), z(t), u(t), p) \quad t \in [0, T] , \tag{5.9a}$$
$$0 = g(x(t), z(t), u(t), p) \quad t \in [0, T] , \tag{5.9b}$$

where t is the time on a horizon $t \in [0, T] \subset \mathbb{R}$, $x \in \mathbb{R}^{n_x}$ are the differential states, $z \in \mathbb{R}^{n_z}$ are the algebraic states, $u \in \mathbb{R}^{n_u}$ are the controls and $p \in \mathbb{R}^{n_p}$ are the time-independent system parameters.

Optimal control problems (OCPs) are mathematical optimisation problems which seek for an optimal process trajectory $x^*(\cdot)$, $z^*(\cdot)$, an optimal control trajectory $u^*(\cdot)$ and possibly optimal process parameters p^* for a process described (here) by a differential-algebraic equation (DAE) process model on a time horizon $[0, T]$. These trajectories and parameters should minimise or maximise a prescribed performance index $\Phi(x(T), z(T), p)$. At the same time, additional path constraints $c(x(t), z(t), u(t), p) \leq 0$ or terminal point constraints $r(x(T), z(T)) \leq 0$ may be present, which have to be satisfied. For fixed initial values of the differential states, the resulting optimal control problem reads as follows:

$$
\begin{aligned}
\min_{\substack{x(\cdot),z(\cdot),\\ u(\cdot),p}} \quad & \Phi(x(T), z(T), p) && \text{(Objective function)} && (5.10\text{a})\\
\text{s.t.} \quad & \dot{x}(t) = f(x(t), z(t), u(t), p) \quad t \in [0, T], && \text{(Dynamic model)} && (5.10\text{b})\\
& 0 = g(x(t), z(t), u(t), p) \quad t \in [0, T], && && (5.10\text{c})\\
& 0 = x(0) - x_0, && \text{(Fixed initial value)} && (5.10\text{d})\\
& 0 \leq c(x(t), z(t), u(t), p) \quad t \in [0, T], && \text{(Path constraints)} && (5.10\text{e})\\
& 0 \leq r(x(T), z(T)) && \text{(Terminal constraints)} && (5.10\text{f})
\end{aligned}
$$

Problem (5.10) describes a continuous time optimal control problem with one model stage (one set of DAEs), a single objective, and fixed initial values. Problem (5.10) serves as starting problem, which, in the following, is modified for the purposes in this thesis. The objective function $\phi(x(T), z(T), p)$ is formulated as a Mayer term, which means the objective function is evaluated at the end of the time horizon. It would also be possible to use an integral formulation $\phi(x(t), z(t), u(t), p)$ which is called Lagrange term. Since a Mayer term can be reformulated as a Lagrange term and vice versa, this is no restriction of generality to only use the Mayer term. Introductions to the theory of optimal control can be found in, e. g., Bryson and Ho (1975), Betts (2001), Diehl (2011), and Gerdts (2012).

Multi-stage DAE-system

As described in Section 5.1, the operation of an adsorption chiller can be described as a multi-stage process (two stages for the one-bed simple cycle). On each stage, e. g. adsorption or desorption phase, the process is described by a different set of DAEs. Therefore, the time horizon $t \in [0, T] \subset \mathbb{R}$ is partitioned into N_{mos} model stages $s \in \{0, \dots, N_{\text{mos}} - 1\}$ on horizons $[t^{(s)}, t^{(s+1)}]$ satisfying $t^{(0)} = 0$, $t^{(N_{\text{mos}})} = T$. The length of the model stages is described by the vector $h \in \mathbb{R}^{N_{\text{mos}}}$. Adding the length of the model stages to the optimisation problem allows to optimise the duration of adsorption and desorption phase times in the case of an adsorption chiller. The modified part of problem (5.10) reads:

$$\min_{\substack{x(\cdot),z(\cdot),\\u(\cdot),p,h}} \quad \Phi(x(T), z(T), p) \qquad \text{(Objective function)} \qquad (5.11a)$$

$$\text{s.t.} \quad \dot{x}(t) = f^{(s)}(x(t), z(t), u(t), p) \quad t \in [t^{(s)}, t^{(s+1)}] \quad s \in [0, N_{\text{mos}} - 1], \qquad (5.11b)$$

$$0 = g^{(s)}(x(t), z(t), u(t), p) \quad t \in [t^{(s)}, t^{(s+1)}] \quad s \in [0, N_{\text{mos}} - 1], \qquad \text{(Dynamic model)} \qquad (5.11c)$$

Cyclic steady state

The performance of an adsorption chiller depends on the initial states x_0. To avoid the effect of arbitrarily chosen initial states, all designs are compared at cyclic steady state operation, i. e. all states at the beginning of a cycle ($t = 0$) equal the states at the end of the cycle ($t = T$). Thus, Equation (5.10d) is replaced by

$$0 = x(0) - x(t = T), \quad \text{(Cyclic steady state)}. \qquad (5.12a)$$

It has to be noted that the cyclic steady state constraint is only employed for comparison and optimisation of designs. For the purpose of model predictive control, a fixed initial value is used corresponding to the current system state (cf. Section 9.1).

Multi-objective optimisation

Until now, only a single-objective optimal control problem is considered. As described in Section 2.1.3, optimisation of an adsorption chiller is a multi-objective optimisation problem with two objective functions, efficiency and power density:

$$\min_{\substack{x(\cdot),z(\cdot),\\ u(\cdot),p,h}} \quad \begin{matrix} \Phi_1(x(T), z(T), p) \\ \Phi_2(x(T), z(T), p) \end{matrix} \qquad \text{(Objective functions)} \qquad \begin{matrix} (5.13a) \\ (5.13b) \end{matrix}$$

To handle two objectives, the ϵ-constraint method is employed (Haimes et al., 1971). This method converts the multi-objective problem into several single objective problems. To do so, the method only optimises one objective while setting a lower or upper bound to all other constraints. By varying this bound (ϵ-constraint), the Pareto frontier can be explored for all objectives. Thus, the objectives (for a case with two objectives) are rewritten to:

$$\min_{\substack{x(\cdot),z(\cdot),\\ u(\cdot),p,h}} \quad \Phi_1(x(T), z(T), p) \qquad \text{(Objective function)} \qquad (5.14a)$$

$$\text{s.t.} \quad \Phi_2(x(T), z(T), p) \leq \epsilon_{\Phi_2} \qquad (\epsilon\text{ -constraint}) \qquad (5.14b)$$

To explore the Pareto frontier, n_ϵ single-objective optimisations are conducted.

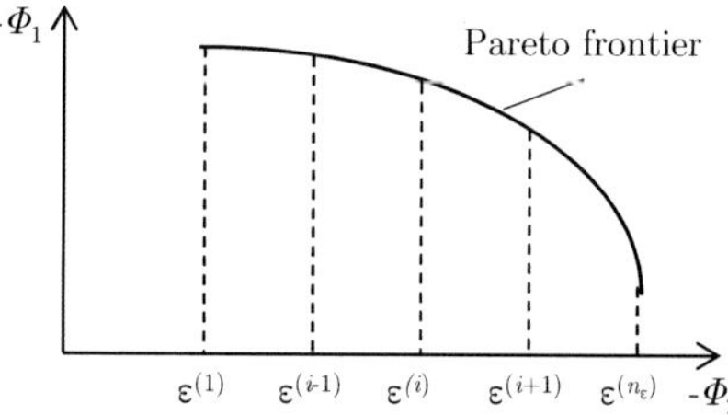

Figure 5.5: Transforming multi-objective optimisation in n_ϵ single-objective optimisations: Pareto frontier and ϵ-constraints.

Resulting optimal control problem

After adding multi-stages, cyclic steady state, and two objective functions, the optimisation problem used for optimising adsorption chiller designs reads:

$$\min_{\substack{x(\cdot),z(\cdot),\\ u(\cdot),p,h}} \quad \Phi_1(x(T), z(T), p) \qquad \text{(Objective function)} \qquad (5.15a)$$

$$
\begin{aligned}
\text{s.t.}\quad \dot{x}(t) &= f^{(s)}(x(t), z(t), u(t), p) \quad t \in [t^{(s)}, t^{(s+1)}] \\
& \qquad s \in [0, N_{\text{mos}} - 1], \qquad \text{(Dynamic model)} \qquad (5.15b) \\
0 &= g^{(s)}(x(t), z(t), u(t), p) \quad t \in [t^{(s)}, t^{(s+1)}] \\
& \qquad s \in [0, N_{\text{mos}} - 1], \qquad (5.15c) \\
0 &= x(0) - x(T), \qquad \text{(Cyclic steady state)} \qquad (5.15d) \\
0 &\leq c^{(s)}(x(t), z(t), u(t), p) \quad t \in [t^{(s)}, t^{(s+1)}] \\
& \qquad s \in [0, N_{\text{mos}} - 1], \qquad \text{(Path constraints)} \qquad (5.15e) \\
0 &\leq \Phi_2(x(T), z(T), p) - \epsilon_{\Phi_2} \qquad (\epsilon\text{ -constraint}) \qquad (5.15f)
\end{aligned}
$$

Figure 5.6 illustrates the differential and algebraic states, the control, the path constraints, the model stages, the cyclic steady state constraint, and the ϵ-constraint. The objective functions to be minimised are $\Phi_1 = -SCP$ and $\Phi_2 = -COP$. The cyclic steady state constraint (5.15d) is a coupled multi-point constraint, connecting the starting point and the end point. The ϵ-constraint (5.15f) is a decoupled point constraint, imposing a constraint at the end point of the time horizon. The path constraints (5.15e) are used to impose the input conditions on the heat exchanger, e. g. $T_{\text{fl,in}} = 30\,°\text{C}$ during adsorption and $T_{\text{fl,in}} = 90\,°\text{C}$ during desorption. In the case of an adsorption chiller, path constrains could additionally be used to limit the heat exchanger outlet temperature or to ensure to ensure a minimum cooling power during the entire time horizon.

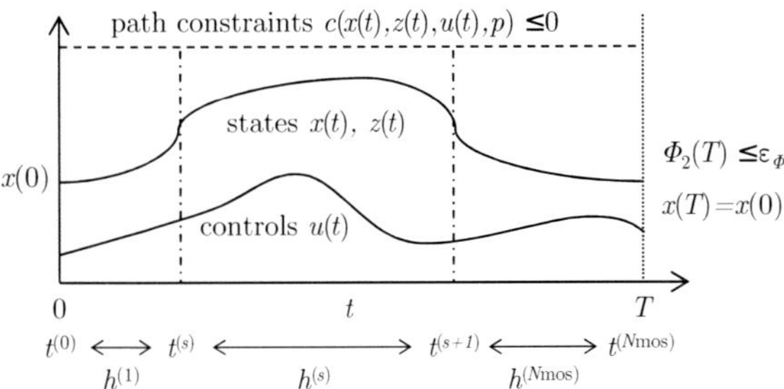

Figure 5.6: Variables and constraints of resulting optimal control problem.

To address continuous time optimal control problems as stated in problem (5.15), three basic approaches exist: (1) state-space, (2) indirect, and (3) direct approaches.

In this thesis, a direct approach is used following the main idea: "first discretise, then optimise". Again, the direct approaches can be classified into three classes: (1)

direct single shooting, (2) direct collocation, and (3) direct multiple shooting. In this thesis, the direct multiple shooting algorithm is used. A good overview and more information on the state-space and the indirect approach for optimal control problems can be found in Diehl (2011).

An implementation of the algorithm is available with the software package MUSCOD-II (Bock and Plitt, 1984; Diehl et al., 2001; Leineweber et al., 2003) that has been used for all computations in this thesis. To connect the dynamic simulation model written in Modelica and the optimisation software package MUSCOD-II, the standardised Functional Mock-up Interface (FMI) is used (Modelica Association Project "FMI", 2014). The dynamic model is encapsulated in a functional mock-up unit (FMU) which contains all differential and algebraic model equations. This FMU is called by the optimisation algorithm. For more details on the technical realisation see Gräber et al. (2012). The full optimisation procedure is shown in Figure 5.7, also indicating the used software implementation (MUSCOD-II), modelling language (Modelica), and interface (FMI).

In the following Sections 5.2.2 and 5.2.3, the used multiple-shooting algorithm is explained in a bit more detail. Since the implementation of the optimisation algorithm itself is not focus of this thesis, the explanations only outline the solution strategy. For further information, the reader is referred to the literature provided at the end of Section 5.2.3.

5.2.2 Direct multiple shooting discretisation and parameterisation

The optimal control problem presented in Section 5.2.1 is an infinite-dimensional optimisation problem. The purpose of the direct multiple shooting method is to transform this problem into a finite-dimensional nonlinear program (NLP) by discretisation of control functions and path constraints, and by parameterisation of state trajectories.

First, a time transformation $t(\hat{t}, h)$ is employed to reformulate the problem on the normalised time horizon $\hat{t} \in [0, N_{\text{mos}}]$ comprising N_{mos} model stages of fixed unit length each,

$$t(\hat{t}, h) := \sum_{i=0}^{s-1} h_i + h_s(\hat{t} - s), \text{ if } \hat{t} \in [s, s+1), \; 0 \leq s \leq N_{\text{mos}} - 1. \tag{5.16}$$

The time transformation is necessary to use the stage durations $h \in \mathbb{R}^{N_{\text{mos}}}$ as subject to optimisation. In the case of an adsorption chiller, the stage durations correspond to the adsorption and desorption phase times which are optimised.

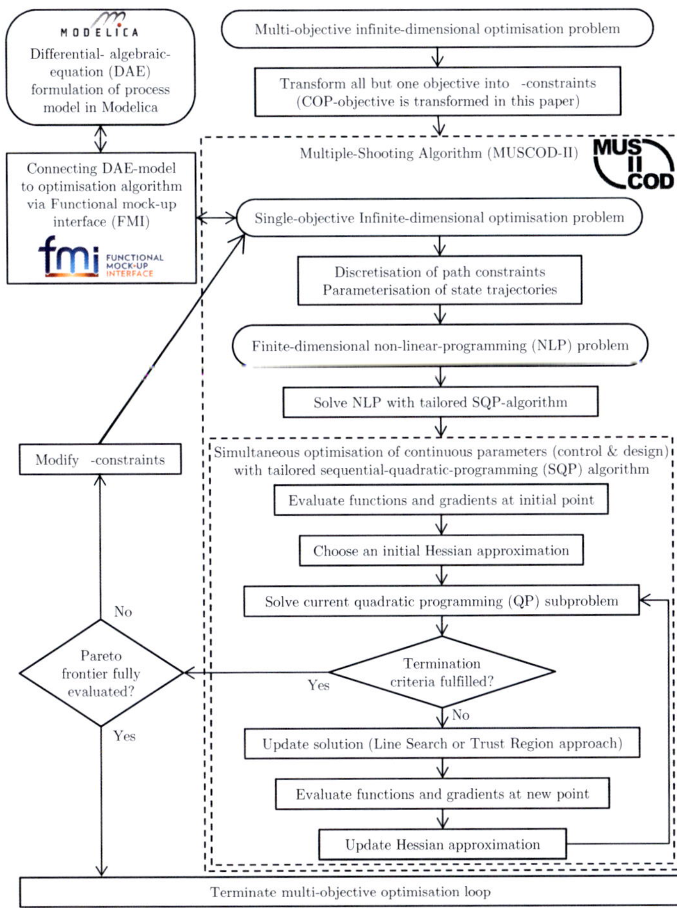

Figure 5.7: Optimisation procedure to evaluate the Pareto frontiers: The ϵ-constraint method is used to generate multiple singe-objective optimisation problems. These problems are solved by a multiple-shooting algorithm which transforms the infinite-dimensional optimisation problem into a NLP problem and solves this problem with a tailored SQP algorithm. Also, the connection between the optimisation algorithm `MUSCOD-II` and the process model written in Modelica via the functional mock-up interface (FMI) is shown.

For discretisation of the states and the control, a shooting grid $\{\hat{t}_i\}$ is introduced on the normalised horizon $[0, N_{\text{mos}}]$,

$$0 = \hat{t}_0 < \hat{t}_1 < \ldots < \hat{t}_{N_{\text{shoot}}} = N_{\text{mos}}. \tag{5.17}$$

comprising all normalised model stages of fixed unit length. For simplicity of exposition, it is assumed that the shooting includes the normalised stage transition points $\hat{t}^{(s)}$, $0 \leq s \leq N_{\text{mos}} - 1$.

Further, the piecewise constant discretisation with control parameters $q_i \in \mathbb{R}^{n_u}$ is introduced,

$$\hat{u}_i(\hat{t}, q_i) := q_i \text{ if } \hat{t} \in [\hat{t}_i, \hat{t}_{i+1}),\ 0 \leq i < N_{\text{shoot}}, \tag{5.18}$$

reducing the control space to functions that can be written as in Equation (5.18) depending on finitely many parameters q_i.

The differential state trajectories are parameterised by multiple shooting state variables s_i on the time grid. The node values serve as initial values for an initial value problem (IVP) solver that computes state trajectories independently on the shooting intervals $\hat{t} \in [\hat{t}_i, \hat{t}_{i+1}]$, $0 \leq i < N_{\text{shoot}}$,

$$\dot{x}_i(\hat{t}) = h_s \cdot f(x_i(t(\hat{t}, h)), \hat{u}_i(t(\hat{t}, h), q_i), p), \quad x_i(\hat{t}_i) = s_i, \tag{5.19a}$$

where $t(\hat{t})$ and h_s depend on the model stage s to which the current shooting interval $[\hat{t}_i, \hat{t}_{i+1}]$ belongs.

For the N_{shoot} state trajectories obtained from these IVPs, continuity is ensured across shooting intervals by introducing additional matching conditions that enter the NLP as equality constraints,

$$s_{i+1} = x_i(t(\hat{t}_{i+1}, h); s_i, q_i, p), \qquad 0 \leq i < N_{\text{shoot}}. \tag{5.20}$$

Here, the solution of the IVP on shooting interval i is denoted by $x_i(t(\hat{t}_{i+1}, h); s_i, q_i, p)$, evaluated in $t(\hat{t}_{i+1}, h)$, and depending on the initial differential states s_i, control parameters q_i, and model parameters p.

5.2.3 Block-structured sequential quadratic programming

From the discretisation and parameterisation described above results a highly structured non-linear program (NLP) of the form

$$\min_{s,q,p,h} \phi(s_{N_{\text{shoot}}}, p) \tag{5.21a}$$

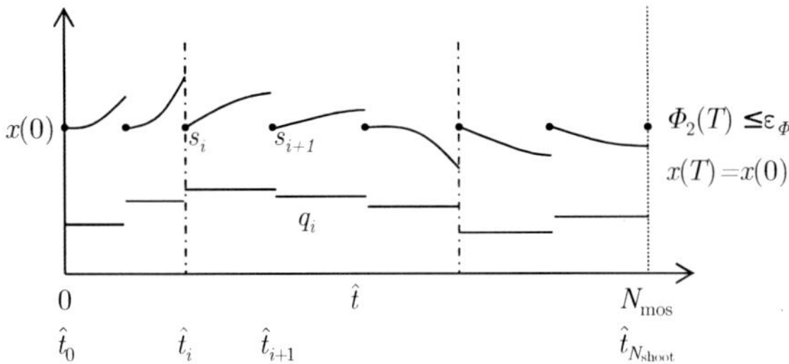

Figure 5.8: Parameterised states s_i, controls q_i, shooting intervals, and model stages in multiple-shooting approach. The figure shows the not converged solution, where continuity is not yet ensured.

$$\text{s.t.} \quad 0 = s_{i+1} - x_i(t(\hat{t}_{i+1}, h); s_i, q_i, p) \quad 0 \leq i \leq N_{\text{shoot}} - 1, \tag{5.21b}$$

$$0 = \sum_{i=1}^{N_{\text{shoot}}} c_i^{\text{eq}}(s_i, q_i, p) \quad 0 \leq i \leq N_{\text{shoot}}, \tag{5.21c}$$

$$0 \leq \sum_{i=1}^{N_{\text{shoot}}} c_i^{\text{in}}(s_i, q_i, p) \quad 0 \leq i \leq N_{\text{shoot}}, \tag{5.21d}$$

where c_i^{eq} comprises a discretisation of all equality constraints contained in problem (5.15) and c_i^{in} comprises all inequality constraints. Additionally, the point constraints to ensure continuity are included in c_i^{eq}.

This large-scale, but structured, NLP is solved by a tailored sequential quadratic programming (SQP) method. This SQP method includes an extensive exploitation of the arising structures. In particular, it uses block-wise high-rank updates of the Hessian approximation, partial reduction techniques for elimination of the algebraic states y_i in the quadratic problems (QP), and condensing techniques for a reduction of the size of this QP to the dimension of the initial values s_0 and controls $(q_0, \ldots, q_{N_{\text{shoot}}-1})$ only.

For more details on the numerical algorithms and techniques employed, the reader is referred to e. g. the textbook Nocedal and Wright (2006) for non-linear programming in general, and to Bock and Plitt (1984) and Leineweber et al. (2003) for details on non-linear programming techniques for the direct multiple shooting.

5.3 Assessing and comparing design choices under optimal control

If only the final optimal design and control parameters are of interest to the designer, it is sufficient to solve the multi-objective optimisation problem once for all continuous design parameters and the control. Nevertheless, it is often worthy to investigate single design parameters to better understand the system. This understanding then enables the designer to derive further improvement measures, probably not yet included into the optimisation problem. Therefore, in this section, the effect of grain size and fin number are assessed individually. First, the optimal control for the reference design is determined. Second, the major heat and mass transfer resistances of this design are identified and optimisation potentials are derived. Third, the effect of grain size is analysed for a given fin number. Fourth, the effect of fin number is investigated for a fixed grain size.

For all optimisations in this chapter, the input parameters (temperatures and volume flow rates) are summarised in Table 5.4. The volume flow rate of the adsorber is chosen according to the value used by Lanzerath (2014) for model calibration.

Table 5.4: Input parameters used for all optimisations in Chapter 5.

Parameter	Value	
$\dot{V}_{\mathrm{fl,A}}$	9	$\mathrm{l\,min^{-1}}$
$T_{\mathrm{fl,in,Des}}$	90	°C
$T_{\mathrm{fl,in,Ads}}$	30	°C
$p_{\mathrm{C}}(30\,°\mathrm{C})$	4.246	kPa
$p_{\mathrm{E}}(10\,°\mathrm{C})$	1.226	kPa

5.3.1 Optimal control for reference design

The optimisation results of the optimal control for the reference design are shown in Figure 5.9: the Pareto frontier and the corresponding time allocations of adsorption and desorption phases for the initial design using a grain size of $d_{\mathrm{p}} = 0.9\,\mathrm{mm}$ and a fin number of $n_{\mathrm{fins}} = 14$.

For this case (reference case, cf. Table 5.2), the highest achievable specific cooling power is $SCP_{\mathrm{ads}} = 327\,\mathrm{W\,kg^{-1}}$ with a corresponding coefficient of performance of

$COP = 0.35$. For the highest achievable COP-value of 0.5, the SCP_{ads} drops to $184\,\mathrm{W\,kg^{-1}}$.

The adsorption and desorption times are continuously rising for higher COP-values. For the maximum SCP_{ads}, the optimal times are $t_{ads} = 243\,\mathrm{s}$ and $t_{des} = 156\,\mathrm{s}$. For the highest COP-value, the times rise to $t_{ads} = 1248\,\mathrm{s}$ and $t_{des} = 453\,\mathrm{s}$. It can be observed that the optimal adsorption time is shorter than the optimal desorption time. However, the phase time ratio does not stay constant for the SCP/COP combinations on the Pareto frontier. Thus, the optimal phase time ratio has to be determined for each operation point.

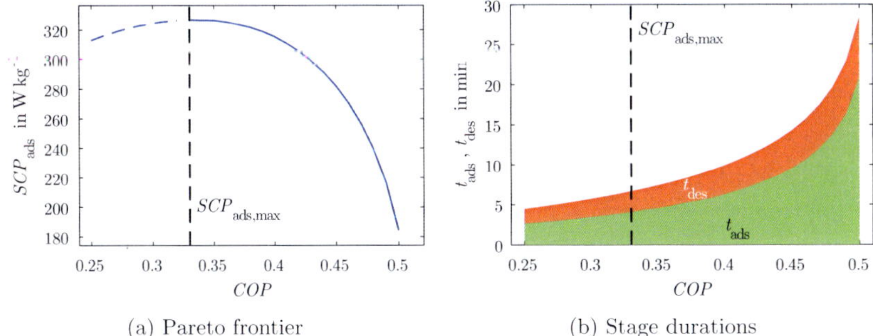

(a) Pareto frontier

(b) Stage durations

Figure 5.9: Results of control optimisation for reference design ($d_p = 0.9\,\mathrm{mm}$, $n_{fins} = 14$). (a) Trade-off between specific cooling power (SCP_{ads}) and coefficient of performance (COP) with Pareto frontier as continuous line. (b) Optimal times for adsorption and desorption corresponding to solutions on Pareto frontier.

5.3.2 Identifying heat and mass transfer resistances by analysing the loss of adsorption potential

In the following, the $SCP_{ads,max}$ solution for the reference design is analysed in detail to identify improvement potential. For this purpose, Figure 5.10 shows the heat flow rates in evaporator and condenser. The heat flow rates have a peak at the beginning of desorption and adsorption. These peaks are characteristic for adsorption chillers and can be explained by using the characteristic curve: Figure 5.11 shows shows the characteristic curve of silica gel 123 and the minimum and maximum achieved filled pore volume (W_{min} and W_{max}). At the beginning of adsorption and desorption phase, the maximum difference in adsorption potential $\Delta A_{ads,max}$ and $\Delta A_{des,max}$ is available.

This maximum difference leads to a steep increase of evaporator and condenser heat flow rate at the beginning of each phase. During each phase, the difference in adsorption potential decreases, which leads to a decrease in driving force and, thus, to a decrease in evaporator and condenser heat flow rate.

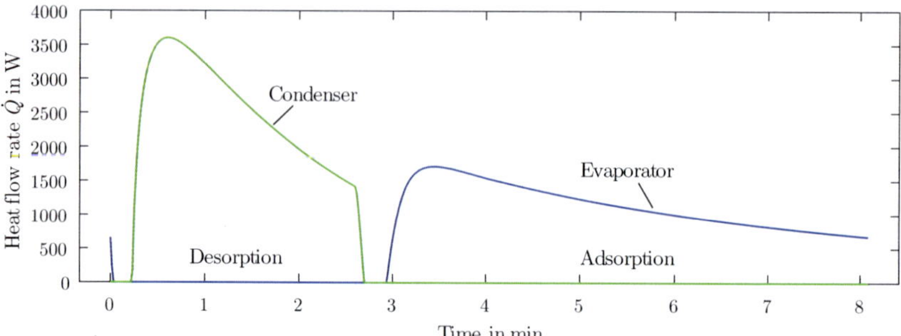

Figure 5.10: Cycle analysis of reference case ($d_p = 0.9\,\text{mm}$, $n_{\text{fins}} = 14$) at $SCP_{\text{ads,max}}$ operation (cf. Figure 5.9). Heat flow rates of evaporator and condenser.

The heat flow rate at the condenser during desorption is higher than the heat flow rate at the evaporator during adsorption. A faster desorption rate is often reported in literature (Aristov et al., 2012b; Graf et al., 2016). In general, three reasons can cause a difference between adsorption and desorption dynamics: (1) difference in heat transfer resistances in the evaporator or condenser, (2) differences in mass transfer resistance for adsorption and desorption, as reported by Lanzerath (2014) for zeolite 13X, and (3) difference in adsorption potential caused by the equilibria data for the given temperature triple.

In the studied system, the evaporator and condenser are ideal components and the mass transfer diffusion coefficient D is equal for adsorption and desorption. Therefore, the higher desorption heat flow rate is caused by the equilibria data. This effect can be nicely illustrated by the characteristic curve (cf. Figure 5.11). Comparing the minimum and maximum differences in adsorption potentials of adsorption and desorption phase $\Delta A_{\text{ads,min}}$, $\Delta A_{\text{des,min}}$, $\Delta A_{\text{ads,max}}$, and $\Delta A_{\text{des,max}}$, it can be observed that $\Delta A_{\text{des,max}} > \Delta A_{\text{ads,max}}$ and, also, $\Delta A_{\text{des,min}} > \Delta A_{\text{ads,min}}$. This larger difference in adsorption potential during desorption leads to a higher driving force during desorption which explains the larger condenser heat flow rate and the optimal adsorption to desorption phase time ratio of $t_{\text{ads}}/t_{\text{des}} > 1$ ($t_{\text{ads}}/t_{\text{des}} = 1.57$).

To improve a given design, it is important to identify the main sources of resistance which are the bottleneck of the design. In the present case study, with ideal evaporator

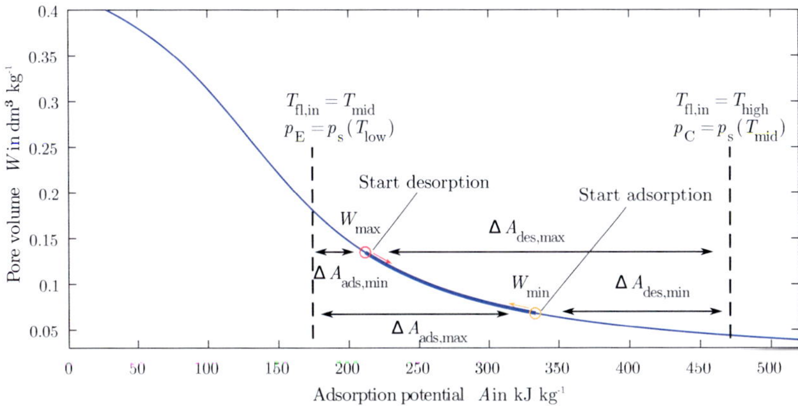

Figure 5.11: Actual minimum and maximum filled pore volume for the $SCP_{\mathrm{ads,max}}$-case of the initial design. Also illustrated are the minimum and maximum adsorption potential at the start and end of adsorption and desorption.

and condenser, five resistances are incorporated in the model: (1) The heat-capacity flow of the heat exchanger fluid $T_{\mathrm{fl,in}} \rightarrow T_{\mathrm{fl}}$. (2) The temperature resistance between fluid and heat exchanger wall $T_{\mathrm{fl}} \rightarrow T_{\mathrm{hx}}$. (3) The temperature resistance between heat exchanger wall and adsorbent: $T_{\mathrm{hx}} \rightarrow T_{\mathrm{ad}}$. (4) The inter-particle mass transfer resistance $p_{\mathrm{E/C}} \rightarrow p^*$. (5) The intra-particle mass transfer resistance $p^* \rightarrow p_{\mathrm{ad}}$. All of these resistance can be translated into a loss in adsorption potential ΔA by calculating the adsorption potential after each resistance $A(T, p)$ (cf. Equation (2.7)).

For the case of adsorption at the moment of maximum heat flow rate, Figure 5.12 shows the loss of adsorption potential caused by the five resistances. It can be observed that the temperature resistance between heat exchanger outer surface and adsorbent material (resistance 3) is a dominant resistance. This resistance can be reduced by increasing the heat exchanger surface A_{sor}, increasing the heat transfer coefficient between heat exchanger and adsorbent α_{sor}, or decreasing the heat capacity of the adsorbent material C_{sor}. The second major resistance in the initial design is the intra-particle mass transfer resistance (resistance 5) which can be decreased by reducing the grain size (cf. Equation (5.5)).

Figure 5.13 (bottom) shows the resistances for desorption and adsorption for the entire cycle. It can be observed that the temperature resistance between heat exchanger outer surface and adsorbent material and the intra-particle mass transfer resistance

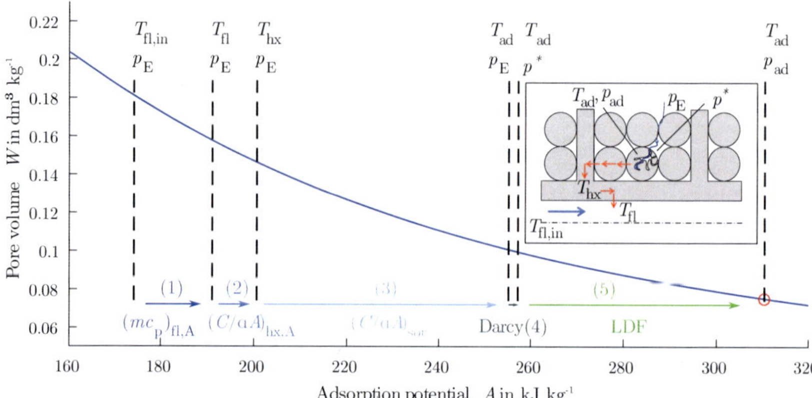

Figure 5.12: Loss of adsorption potential for the reference design at $SCP_{ads,max}$ operation for the moment of maximum cooling power in the evaporator (cf. Figure 5.10). The resistances from left to right are: (1) heat capacity flow of heat exchanger fluid $(\dot{m}c_p)_{fl,A}$, (2) temperature resistance between heat exchanger fluid and heat exchanger wall $(C/\alpha A)_{hx,A}$, (3) temperature resistance between heat exchanger wall and adsorbent $(C/\alpha A)_{sor}$, (4) inter-particle mass-transfer resistance (Darcy), (5) and intra-particle mass-transfer resistance (LDF).

are dominant throughout the cycle. In the following, this illustration of the loss in adsorption potential is used to analyse and compare improved designs.

5.3.3 Sensitivity analysis of design choices – grain size effect

In this section, the effect of the grain size on performance is investigated under optimal control. In general, changing the grain size may affect various model parameters. The grain size directly affects the mass transfer coefficients β_{Darcy} and β_{LDF}, which is reflected in equations (5.3) and (5.5). It may also have some effects on the effective heat transfer coefficient α_{ad}, the heat exchanger surface to adsorbent mass ratio A/m_{sor}, and the heat capacity ratio C_{hx}/C_{sor} due to changes in contact resistance and adsorbent mass.

In this thesis, uniformly-sized perfect spheres are assumed, for which porosity can be considered to be independent of particle diameter (Ouchlyama and Tanaka, 1984). Thus, although size and distribution of the pores alter by changing the grain size, their global share of volume within the adsorbent layer remains constant, leaving

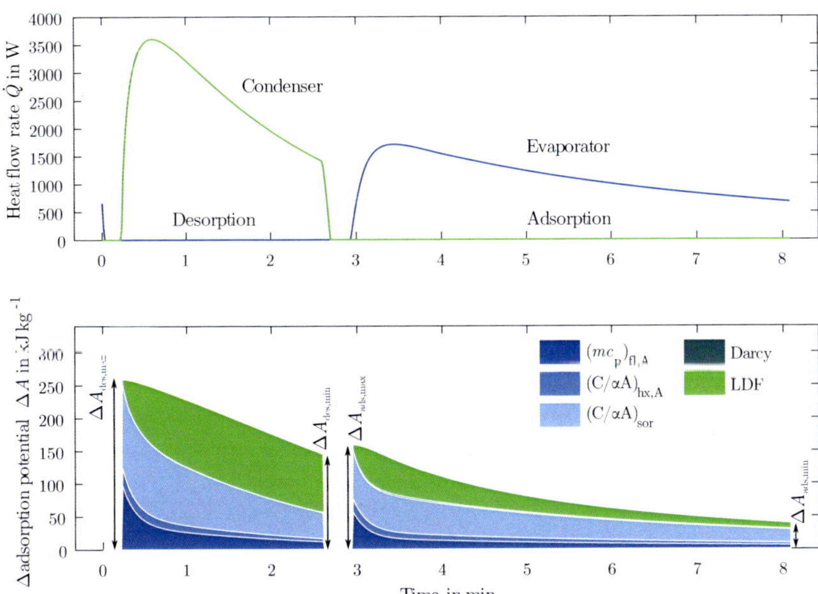

Figure 5.13: Analysis of reference case ($d_\mathrm{p} = 0.9\,\mathrm{mm}$, $n_\mathrm{fins} = 14$) at $SCP_\mathrm{ads,max}$ operation (cf. Figure 5.9). (top) Heat flow rates of evaporator and condenser. (bottom) Loss of adsorption potential ΔA due to heat and mass transfer resistances (cf. Figure 5.12).

the adsorbent mass constant as well. Based on the Zehner-Schlunder model (Hsu et al., 1994), the equivalent thermal conductivity of an adsorbent layer $\lambda_\mathrm{ad,eq}$, is only a function of adsorbent layer porosity ϵ_b and the specific thermal conductivity of the working pair. Since working pair and porosity do not change, it is also assumed that the effective heat transfer coefficient α_ad is constant, since it is a function of the adsorbent layer equivalent thermal conductivity $\lambda_\mathrm{ad,eq}$.

In terms of mass transfer, however, changing the grain size has two opposing effects on the inter-particle (Darcy) and intra-particle (LDF) mass transfer resistances in both adsorption and desorption phases:

For large grain sizes, the inter-particle mass transfer resistance has a minor contribution to the mass transfer resistance because of the increased pore volume between the particles. Instead, large grains suffer from high intra-particle mass transfer resistance, since adsorbate vapour has to travel longer distances within the particle. In this so-called intra-particle mass transfer dominated regime, the LDF approach is preferred for describing mass diffusion within the adsorbent layer.

For very small grains, in contrast, inter-particle mass transfer resistance becomes the dominating resistance because of the reduced permeability caused by small grains. This resistance causes a high pressure gradient along the adsorbent layer. Due to a small particle diameter, the intra-particle resistance has a minor contribution to the global mass transfer resistance in this range, and Darcy's law can be used to describe the mass transfer behaviour.

However, in a certain grain size range, none of the resistances dominates the other one. The global mass transfer resistance finds its lowest value in this range, and the heat transfer resistance may become dominating. Aristov (2013b) called this range "lumped mode" or "grain size insensitive regime" since a variation of grain size within this range will not effect the system's dynamics and performance.

Figure 5.14 illustrates the contribution of inter-particle and intra-particle mass transfer resistances to the overall pressure loss. The figure shows the average pressure differences during adsorption and desorption phase for different grain sizes at points of maximum cooling power. The overall mass transfer resistance is minimal for particle diameters around $d_\mathrm{p} = 0.3\,\mathrm{mm}$ to $0.75\,\mathrm{mm}$ for adsorption and $d_\mathrm{p} = 0.14\,\mathrm{mm}$ to $0.35\,\mathrm{mm}$ for desorption: in these ranges, the total pressure differences do not exceed the minimum total pressure differences by 20 %.

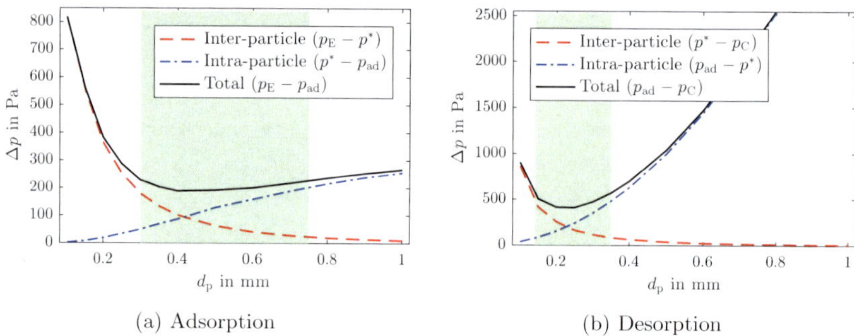

(a) Adsorption

(b) Desorption

Figure 5.14: Average pressure differences for different grain sizes at points of maximum cooling power during (a) adsorption and (b) desorption. Total pressure difference (black –) is caused by inter-particle share (red – –) and intra-particle share (blue .–). Lumped mode is indicated by the green shaded area.

Figure 5.15 shows the Pareto frontiers for grain sizes from $d_p = 0.1$ mm to 1 mm. For the smallest investigated grain size $d_p = 0.1$ mm, the adsorber-bed's specific cooling power does not exceed $SCP_{ads} = 240\,\mathrm{W\,kg^{-1}}$. The performance increases for larger grains. Here, the system is in the inter-particle dominated regime. The highest achievable specific cooling power is $SCP_{ads} = 453\,\mathrm{W\,kg^{-1}}$ for a grain size of $d_p = 0.35$ mm. This is an increase in $SCP_{ads,max}$ of 38.5 % compared to the reference grain size $d_p = 0.9$ mm. The optimal grain size is found in the "lumped mode" for grain sizes of $d_p = 0.25$ mm to 0.5 mm, where the performance does not vary much. These grain sizes provide the highest performance in terms of both COP and SCP_{ads}. By further increasing the grain size, the Pareto frontier traverses back towards lower SCP_{ads} and COP values.

It should be emphasised that all adsorbers are operated under optimal control as determined from the optimisation problem. The optimisation of control allows to compare adsorber beds with different particle diameters, since each design is assessed at its best operating point, i.e., stage durations which fit the design's dynamic characteristics.

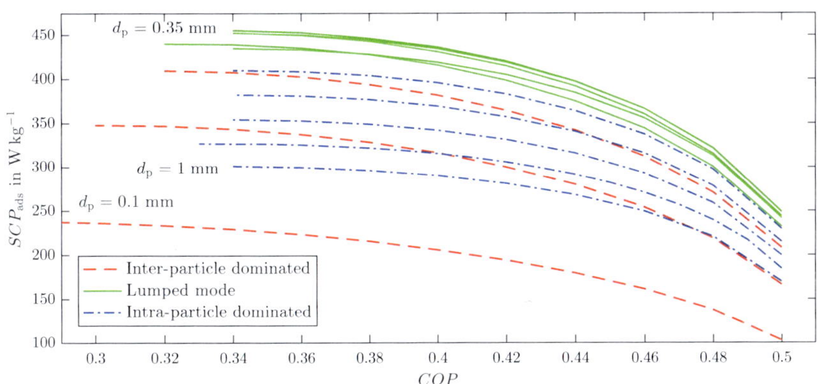

Figure 5.15: Pareto frontiers regarding SCP_{ads} and COP for different grain sizes: inter-particle dominated regime $d_p = 0.1$ mm to 0.25 mm (red −−), lumped mode $d_p = 0.25$ mm to 0.5 mm (green −) and intra-particle dominated regime $d_p = 0.5$ mm to 1 mm (blue (.−)).

5.3.4 Sensitivity analysis of design choices – number of fins

As a second example for rigorous assessment and comparison of adsorber designs, the sensitivity analysis of design choices, the number of fins is varied for the given heat exchanger (see Figure 5.3a) in the model-based analysis. Changing the number of fins has three opposing effects on the model parameters and accordingly on the objective functions: (1) An increasing number of fins directly increases the metal mass which has to be cooled and heated. Thus, the heat capacity ratio $C_{\mathrm{hx}}/C_{\mathrm{sor}}$ increases and the COP drops. (2) At the same time, the heat exchanger surface area A increases by the additional fins which increases the ratio of heat exchanger surface to adsorbent mass $A_{\mathrm{ad}}/m_{\mathrm{sor}}$. This leads to an improved heat transfer and a higher SCP_{ads}. (3) Increasing the fin number also decreases the mass of adsorbent within the adsorber bed, since the available space is reduced. A reduced adsorbent mass again has multiple effects on COP and SCP_{ads}: First, the heat capacity ratio $C_{\mathrm{hx}}/C_{\mathrm{sor}}$ is further increased and also less adsorbent is adsorbed both leading to a reduced COP. Second, the adsorbent mass is in the denominator of the specific cooling power SCP_{ads} which, therefore, increases this objective function.

Designs with fin numbers ranging from $n_{\mathrm{fins}} = 6$ to 22 fins per tube have been analysed. The effect of fin number on the system parameters is captured in equations (5.6) - (5.8). The particle diameter is kept constant at $d_{\mathrm{p}} = 0.35\,\mathrm{mm}$. The control is again optimised for each design leading to a design-specific Pareto frontier (Figure 5.16). It can be observed that by increasing the number of fins, the Pareto frontier of the design shifts to higher SCP_{ads} and lower COP-values. For a fin number of $n_{\mathrm{fins}} = 22$, the maximum SCP_{ads} can be increased by 43 % to $SCP_{\mathrm{ads}} = 652\,\mathrm{W\,kg^{-1}}$ compared to the case with $n_{\mathrm{fins}} = 14$ and a particle diameter of $d_{\mathrm{p}} = 0.35\,\mathrm{mm}$. Compared to the reference case ($n_{\mathrm{Fins}} = 14$, $d_{\mathrm{p}} = 0.9\,\mathrm{mm}$, $SCP_{\mathrm{ads,max}} = 327\,\mathrm{W\,kg^{-1}}$), the maximum specific cooling power could almost be doubled.

5.4 Simultaneous optimisation of grain size, number of fins, and control

If the designer is not interested in analysing and quantifying the effects of certain design choices on COP and SCP_{ads}, but wants to find the optimal design, a simultaneous optimisation of design and control should be conducted. The optimisation problem is, therefore, extended by the design parameters, which are to be optimised, in this case grain size and fin number. It has to be noted that, strictly speaking, the fin

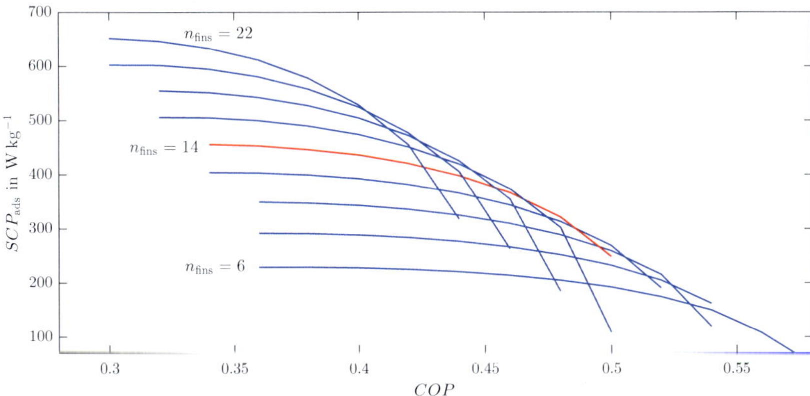

Figure 5.16: Pareto frontiers regarding SCP_{ads} and COP for varying number of fins on the adsorber tube. The initial case with a fin number of $n_{\text{fins}} = 14$ is indicated as red line.

number is a discrete optimisation parameter. In contrast to a working pair or a cycle design, however, the fin number can be easily relaxed to a continuous design parameter since all model equations containing the fin number allow for non-integer variables (cf. Equations 5.6 - 5.8). Therefore, in the following, the fin number is used as a continuous design parameter.

The resulting Pareto frontier is shown in Figure 5.17a. For comparison, the Pareto frontiers obtained for changing fin numbers (cf. Figure 5.16) are also plotted. It can be observed that the overall Pareto frontier is the envelope to the Pareto frontiers of the sub-problem of changing fin numbers discussed in Section 5.3.4. Figure 5.17b shows the corresponding design parameters. The optimal particle diameter stays in a narrow range between $d_{\text{p}} = 0.3\,\text{mm}$ to $0.4\,\text{mm}$ in the lumped mode. The optimal fin number has its maximum value $n_{\text{fins}} = 22$ for COPs up to 0.39 and then successively decreases when moving to higher COPs. Given the results from Section 5.3.3 and 5.3.4, these results could be expected and prove consistency of the method.

To compare the optimised design ($n_{\text{fins}} = 22$ and $d_{\text{p}} = 0.325\,\text{mm}$) with the initial design $n_{\text{fins}} = 14$ and $d_{\text{p}} = 0.9\,\text{mm}$, Figure 5.18 shows the heat flow rates and the loss of adsorption potential for both designs at $SCP_{\text{ads,max}}$ operation. Compared to the initial design (Figure 5.13) four observations can be made: (1) the overall heat flow rate increased during both condensation and vaporisation. This is caused by a reduced overall resistance and leads to shorter adsorption and desorption phase

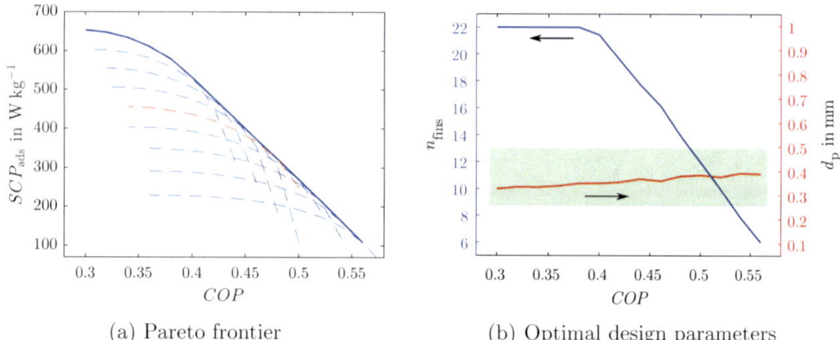

(a) Pareto frontier

(b) Optimal design parameters

Figure 5.17: Results of the simultaneous optimisation of design (grain size and fin number) and control. (a) Overall Pareto frontier obtained by simultaneous optimisation (bold line). For comparison, Pareto frontiers of specific fin numbers (dashed lines, cf. Figure 5.16). (b) Optimised design parameters, grain size and fin number, corresponding to solutions on the overall Pareto frontier.

times and to a higher specific cooling power. (2) The optimised design shows reduced resistance of intra-particle mass transfer (LDF) but a higher resistance of inter-particle mass transfer (Darcy). This effect can be attributed to the reduced particle diameter. (3) The share of overall heat transfer resistance increases compared to overall mass transfer resistance (blue area compared to green area in Figure 5.18). (4) Within the adsorber heat exchanger, the resistance also shifted. The share of temperature resistance between heat exchanger and adsorbent decreased compared to the share of the heat capacity flow rate and the inner temperature resistance within the tube.

From this analysis, the next main bottlenecks can be derived: the heat capacity flow rate in the adsorber heat exchanger and, still, the heat transfer from the heat exchanger to the adsorbent. The first resistance (heat capacity flow rate) could be reduced by a higher mass flow rate in the heat exchanger which would cause a higher electricity demand for pumping in the adsorption chiller. This trade-off would lead to the next optimisation problem. The second resistance (heat transfer from heat exchanger to adsorbent) could be increased by a further increase in heat transfer area or by an increase in the specific heat transfer coefficient. In practice, a higher specific thermal coefficient between heat exchanger and adsorber could be achieved by a coating (Restuccia et al., 2002). Since a coating would change the ratio between heat exchanger mass and adsorbent mass, a reduced COP is expected. To decide which option is preferable, a full Pareto frontier is needed including these design choices.

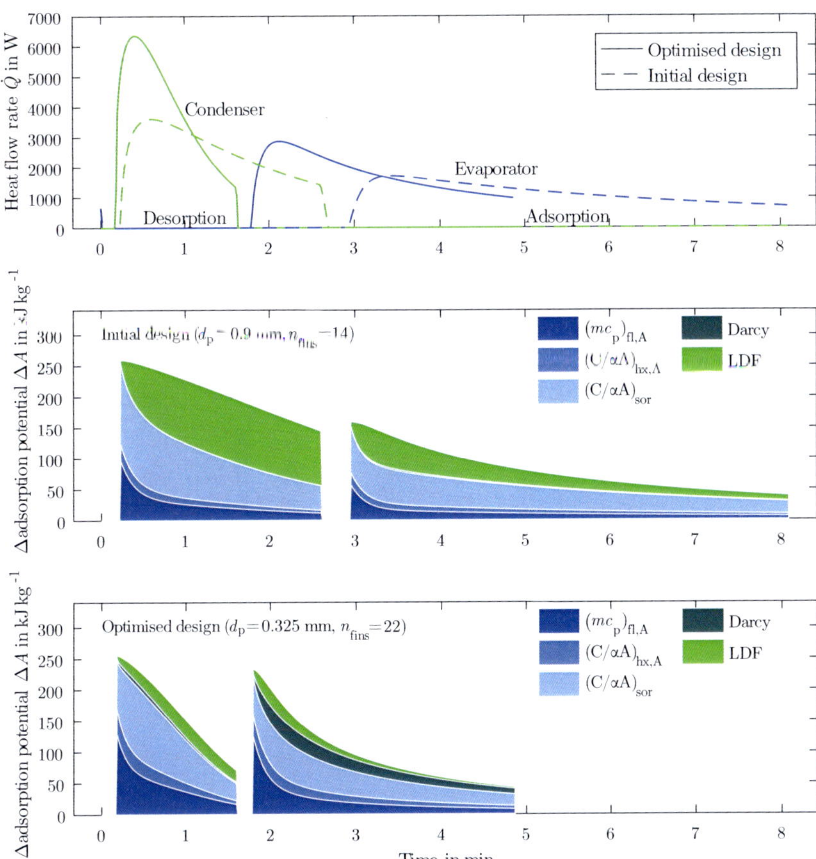

Figure 5.18: Comparison of initial (d_p = 0.9 mm, n_{fins} = 14) and optimised design (d_p = 0.325 mm, n_{fins} = 22) using the loss of adsorption potential at $SCP_{ads,max}$ operation (cf. Figure 5.17). (top) Heat flow rates of evaporator and condenser. (Middle) Initial design: loss of adsorption potential ΔA due to heat and mass transfer resistances (cf. Figure 5.12). (bottom) Optimised design: loss of adsorption potential ΔA due to heat and mass transfer resistances.

In this chapter, the adsorber bed of a finned-tube heat exchanger with loose silica gel grains has been assessed and optimised. The simultaneous optimisation of grain size, fin number, and control has been conducted for a simple one-bed cycle and ideal evaporator and condenser characteristics. As objective functions, the specific cooling power with respect to the adsorbent mass SCP_{ads} and the coefficient of performance COP have been used. The simultaneous optimisation of design and control is further used in Chapters 6 - 8. In the next chapter, the investigated system is extended by real evaporator and condenser components, and the effect of a changed measure for the specific cooling power on optimal design is investigated.

CHAPTER 6

Simultaneous optimisation of adsorber-bed design, component sizing, and control

In the last chapter, the idea of rigorous model-based assessment and optimisation of adsorption chillers has been introduced and exemplified. The focus in the first example has been the adsorber bed in a one-bed cycle with ideal evaporator and condenser components. In this chapter, a complete adsorption chiller is investigated including evaporator and condenser: First, the dynamic model is enhanced by non-ideal evaporator and condenser models and the performance reduction is quantified, which is caused by the additional resistances (Section 6.1). To optimise the sizing of evaporator and condenser, the objective function SCP_{ads} is unsuitable, since it does not penalise the components' weights. Therefore, a new measure for power density is introduced which takes into account the entire system mass SCP_{all}. The optimisation regarding both measures for power density (SCP_{ads} and SCP_{all}) is compared in Section 6.2. In Section 6.3, a stepwise optimisation of adsorber design and evaporator and condenser sizing is conducted to separate the effect of each optimisation on the performance. Figure 6.1 shows the discrete design choices and optimisation parameters considered in this chapter.

6.1 Enhancement of dynamic model and performance of initial design

In this section, the ideal pressure boundaries used in the initial model (Chapter 5) are replaced by more complex models for evaporator and condenser. Both models

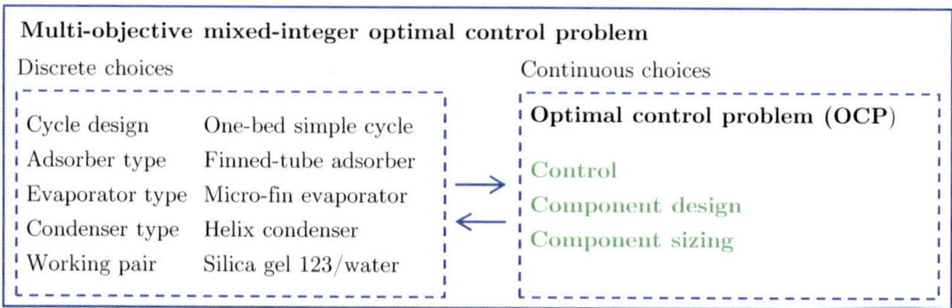

Figure 6.1: Discrete choices and considered optimisation parameters (green) for simultaneous optimisation of adsorber-bed design, component sizing, and control: optimisation of finned-tube adsorber, micro-fin evaporator, helix condenser, and one-bed simple cycle.

consist of three cell elements each connected by two heat transfer models. The three cell elements are the heat exchanger liquid, the heat capacity of the heat exchanger tube, and the working fluid in the two-phase region (cf. Chapter 3). The heat transfer models connect these cells by corresponding heat transfer coefficients αA. Additionally, a return valve is added to the model which regulates the liquid flow between condenser and evaporator. The new model has 12 differential states in total. Figure 6.2 illustrates the model structure, including all states.

With this model structure, three resistances are considered in each heat exchanger: (1) the heat capacity flow of the heat exchanger fluid, (2) the temperature resistance between the heat exchanger fluid and the heat exchanger metal, and (3) the temperature resistance between the heat exchanger metal and the working fluid.

The evaporator and condenser models are parametrised by geometric data and correlations for the heat transfer coefficients. In this thesis, a micro-fin evaporator and a helix-shaped condenser are used. Figure 6.3 shows a picture of the heat exchangers. In both heat exchangers, the heat exchanger fluid is flowing inside the heat exchanger tube and the working fluid is vaporised/condensed at the outside of the heat exchanger tube.

Both heat exchangers have been experimentally characterised in our lab by Lanzerath et al. (2015), so that the heat transfer coefficients are known. For the evaporator, the heat transfer between heat exchanger fluid and heat exchanger metal is calculated with the Sieder-Tate correlation (Sieder and Tate, 1936). The heat transfer between heat exchanger metal and working pair is assumed to be constant and is experimentally determined by Lanzerath et al. (2015). For the condenser, the heat transfer between

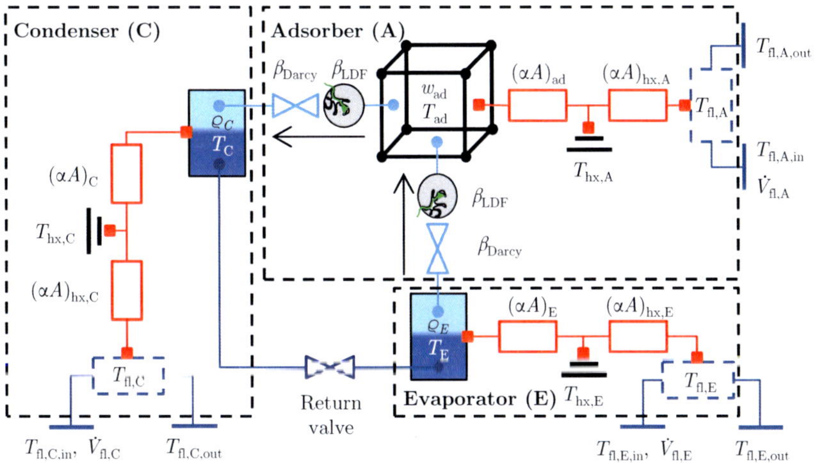

Figure 6.2: Structure of the dynamic adsorber–bed model including non-ideal models of evaporator and condenser.

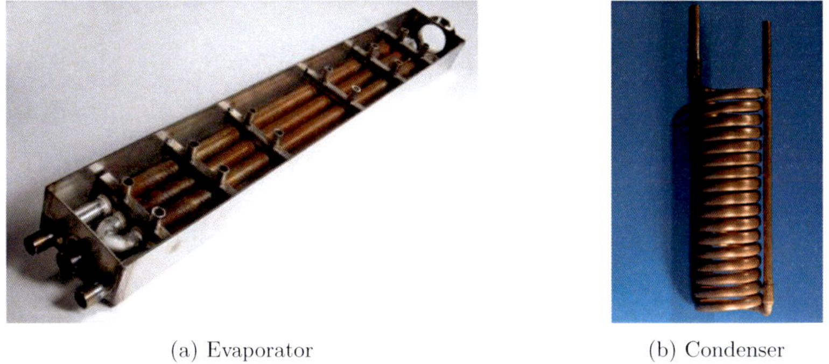

(a) Evaporator (b) Condenser

Figure 6.3: Picture of evaporator and condenser heat exchanger.

heat exchanger fluid and heat exchanger metal is calculated with the Schmidt (1967) correlation, which takes into account the effect of centrifugal forces on the heat transfer induced by the helix design. Again, the heat transfer between heat exchanger metal and working pair is experimentally determined by Lanzerath et al. (2015). All parameters are listed in Table 6.1.

Table 6.1: Parameters of evaporator and condenser determined by Lanzerath et al. (2015).

Parameter	Value	
V_{C}	2	l
V_{E}	2	l
$(\alpha A)_{\mathrm{hx,C}}/l_{\mathrm{hx,C}}$	3.89×10^2	$\mathrm{W\,K^{-1}\,m^{-1}}$
$(\alpha A)_{\mathrm{hx,E}}/l_{\mathrm{hx,E}}$	2.04×10^2	$\mathrm{W\,K^{-1}\,m^{-1}}$
$(\alpha A)_{\mathrm{C}}/l_{\mathrm{hx,C}}$	6.98×10^2	$\mathrm{W\,K^{-1}\,m^{-1}}$
$(\alpha A)_{\mathrm{E}}/l_{\mathrm{hx,E}}$	7.62×10^1	$\mathrm{W\,K^{-1}\,m^{-1}}$
$C_{\mathrm{hx,C}}/l_{\mathrm{hx,C}}$	9.63×10^1	$\mathrm{J\,K^{-1}\,m^{-1}}$
$C_{\mathrm{hx,E}}/l_{\mathrm{hx,E}}$	2.27×10^2	$\mathrm{J\,K^{-1}\,m^{-1}}$

For all following optimisations in the Chapters 6 - 8, the volume flow rates are chosen according to the value used by Lanzerath et al. (2015) for model calibration and are summarised in Table 6.2. The used input temperatures are stated specifically for each optimisation.

Table 6.2: Input parameters for volume flow rates used for all optimisations in in the Chapters 6 - 8.

Parameter	Value	
$\dot{V}_{\mathrm{fl,A}}$	9	$\mathrm{l\,min^{-1}}$
$\dot{V}_{\mathrm{fl,C}}$	7	$\mathrm{l\,min^{-1}}$
$\dot{V}_{\mathrm{fl,E}}$	12	$\mathrm{l\,min^{-1}}$

Performance of initial design

To analyse the performance reduction caused by the evaporator and condenser, the Pareto frontier for optimal control is determined by solving the optimisation problem described in Section 5.2. The length of evaporator and condenser is chosen in accordance to the components used by Lanzerath (2014).

Figure 6.4 (a) shows the Pareto frontiers for the reference design with ideal evaporator/condenser (blue) and real evaporator/condenser (red). The maximum achievable SCP_{ads} drops by 38 % from $SCP_{\mathrm{ads}} = 326\,\mathrm{W\,kg^{-1}}$ to $SCP_{\mathrm{ads}} = 200\,\mathrm{W\,kg^{-1}}$. The maximum achievable COP is not influenced, because it is only a function of working pair and adsorber-bed design (cf. Section 2.1.3). Figure 6.4 (b1) shows the total cycle time and Figure 6.4 (b2) shows the ratio between adsorption and desorption phase time. The overall cycle time increases when switching from ideal to real components, due to the added resistances. Also, the ratio between adsorption and desorption shifts to higher adsorption times. This is an indication for higher resistances in the evaporator compared to the condenser. For solutions with a high SCP, the optimal phase time ratio scatters slightly. This is caused by the termination criteria of the optimisation: the optimisation terminates when a given tolerance for the objective function (SCP) is achieved. The scattering of the optimal cycle times lie within this tolerance and, thus, the scattering is not observable for the Pareto frontier.

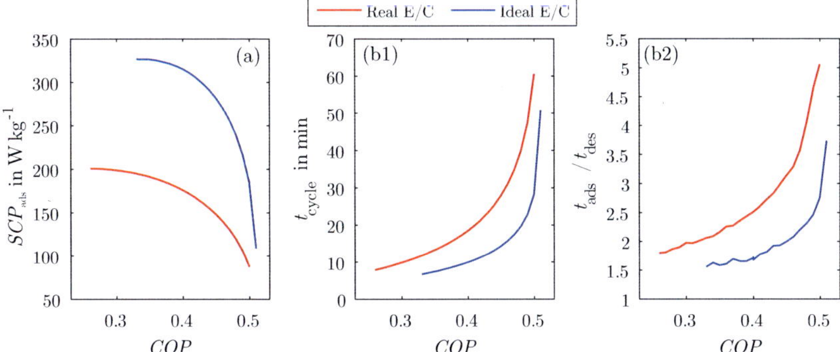

Figure 6.4: Comparison of ideal evaporator/condenser (blue) and real evaporator/condenser (red) for initial adsorber design: (a) Trade-off between specific cooling power (SCP_{ads}) and coefficient of performance (COP). (b1) Optimal overall cycle times ($t_{\text{ads}} + t_{\text{des}}$). (b2) Optimal adsorption / desorption phase time ratio ($t_{\text{ads}}/t_{\text{des}}$).

Figure 6.5 shows heat flow rates (top) and differences in adsorption potential distinguished by resistances (bottom) of the reference case with real evaporator and condenser. Compared to the reference case with ideal evaporator and condenser (Figure 5.13), the additional resistances of evaporator and condenser are shown in red. During desorption, the additional resistance is caused by the condenser; during adsorption, the additional resistance is caused by the evaporator. The resistances of the evaporator during adsorption have a higher share compared to the resistances of the condenser during desorption. This additional resistances explain the increased adsorption/desorption time ratio. The highest resistance in the evaporator is the heat transfer resistance on the outside of the tubes. To reduce this heat transfer resistance

Lanzerath et al. (2016) proposed coated evaporator tubes.

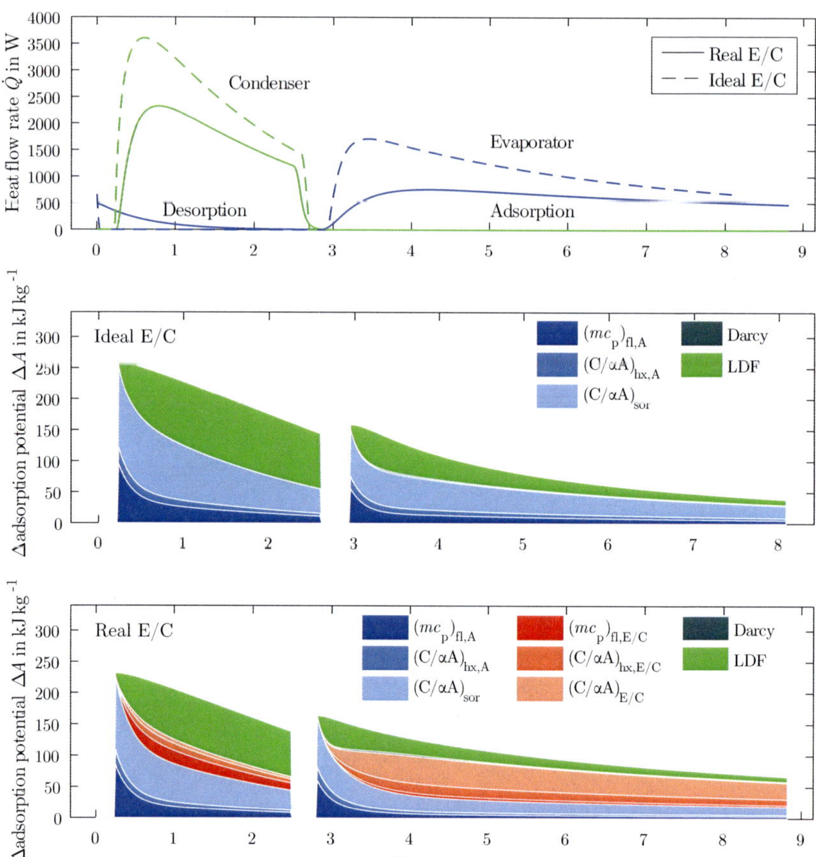

Figure 6.5: Comparison of initial design with ideal evaporator and condenser and initial design with real evaporator and condenser using the loss of adsorption potential at $SCP_{\mathrm{ads,max}}$ operation (cf. Figure 5.17). (top) Heat flow rates of evaporator and condenser. (middle) Ideal evaporator and condenser: loss of adsorption potential ΔA due to heat and mass transfer resistances (cf. Figure 5.12). (bottom) Real evaporator and condenser: loss of adsorption potential ΔA due to heat and mass transfer resistances: additional resistances in evaporator and condenser.

The next step is the optimal sizing of evaporator and condenser to find a more balanced adsorption chiller design. This is not possible with the objective function SCP_{ads}, since it does not penalise the weight of evaporator and condenser. Thus, an unconstrained optimisation would increase evaporator and condenser infinitely to reduce heat transfer resistances. Therefore, a new measure for power density SCP_{all} is introduced in the next section and the influences on optimal design are analysed.

6.2 Sensitivity to objective function

In Section 2.1.3, different reference measures for the specific cooling power are introduced, such as the adsorbent mass, the mass of all heat exchangers or the volume of the entire adsorption chiller. Until now, the most common measure, the adsorbent mass has been used:

$$SCP_{\text{ads}} = \frac{\bar{\dot{Q}}_{\text{E}}(n_{\text{fins}}, d_{\text{p}}, l_{\text{hx,E}}, l_{\text{hx,C}})}{m_{\text{sor}}(n_{\text{fins}})} . \tag{6.1}$$

When using SCP_{ads} as objective function and including the sizing of evaporator and condenser into the optimisation, the optimal solution will presumably be the maximum allowed size of evaporator and condenser. Therefore, it is necessary to include the mass of all heat exchangers, leading to:

$$SCP_{\text{all}} = \frac{\bar{\dot{Q}}_{\text{E}}(n_{\text{fins}}, d_{\text{p}}, l_{\text{hx,E}}, l_{\text{hx,C}})}{m_{\text{sor}}(n_{\text{fins}}) + m_{\text{hx,ads}}(n_{\text{fins}}) + m_{\text{hx,E}}(l_{\text{hx,E}}) + m_{\text{hx,C}}(l_{\text{hx,C}})} . \tag{6.2}$$

The dependencies of cooling power and mass on the design parameters are shown explicitly in Equation (6.2) to highlight the importance of including all masses.

In the following, two optimisations are conducted. The first optimises design and control regarding COP and SCP_{ads}. The second optimises design and control regarding COP and SCP_{all}.

Figure 6.6 shows the optimisation results regarding the two objective functions SCP_{ads} and SCP_{all} and all optimised parameters (t_{cycle}, $^{t_{\text{ads}}}/_{t_{\text{des}}}$, d_{p}, n_{fins}, $l_{\text{hx,E}}$ and $l_{\text{hx,C}}$). As upper bounds for the evaporator and condenser length, 5 times the initial length is chosen: $l_{\text{hx,E,max}} = 11.55\,\text{m}$ and $l_{\text{hx,C,max}} = 22.75\,\text{m}$.

Figure 6.6 (a1) and (a2) show the achieved SCP_{ads}/COP and SCP_{all}/COP for both optimisations. As expected, higher SCP_{ads}/COP are achieved when SCP_{ads} is used as objective, and vice versa. When changing the objective function form SCP_{ads} to SCP_{all}, the maximum SCP_{ads} drops by 44 % from $490.7\,\text{W}\,\text{kg}^{-1}$ to $273.4\,\text{W}\,\text{kg}^{-1}$, but maximum SCP_{all} increases by 59 % from 41.7 W kg^{-1} to 66.2 W kg^{-1}. These first results show the importance of the objective function.

As expected, the optimisation with SCP_{ads} as objective function sets both heat exchanger lengths to their maximum value (Figure 6.6 (c3) and (c4)). The optimisation regarding SCP_{all} including heat exchanger mass gives entirely different results: The optimal evaporator heat exchanger length is around $l_{\text{hx,E}} \approx 3.5$ m, which is 52 % larger compared to the reference design. The optimal condenser heat exchanger length is around $l_{\text{hx,C}} \approx 2.1$ m, which is around half the size of the reference design. The optimal fin number significantly decreases compared to optimisation regarding SCP_{ads} (Figure 6.6 (c2)): for SCP_{max} operation the optimal fin number decreases from $n_{\text{fins}} = 22$ to $n_{\text{fins}} \approx 15$. Optimal cycle times increase, while optimal ratios of adsorption/desorption phase time decrease (Figure 6.6 (b1) and (b2)). Only the optimal particle diameter is almost independent of the chosen objective function. For both cases, the optimal particle diameter is around $d_{\text{p}} \approx 0.4$ mm, which lies in the "grain size insensitive regime".

The effects on design and control can be explained as follows:

Reduced heat exchanger length of evaporator and condenser: The length of the heat exchangers are a direct cost when using SCP_{all} as objective function. Therefore, the optimisation seeks for a compromise between increasing cooling power and increasing heat exchanger mass.

Reduced optimal fin number in adsorber: The reduced fin number is caused by two effects: First, the mass of adsorber heat exchanger is now also incorporated into the objective function SCP_{all}. Since the density of the aluminium fins is higher than that of the adsorbent material, a smaller number of fins leads to a lower weight of the adsorber heat exchanger. Second, due to the higher heat transfer resistances in evaporator and condenser heat exchanger, a lower number of fins in the adsorber heat exchanger is sufficient to achieve a balanced design.

Increased optimal cycle time: The smaller evaporator and condenser heat exchangers cause a higher resistance in evaporator and condenser and, thus, lead to longer optimal cycle times.

Decreased optimal adsorption/desorption phase time ratio: The sizes of evaporator and condenser heat exchanger decrease compared to the reference design. Therefore, the ratio of resistances in evaporator and condenser shifts to a relatively higher resistance in the condenser. This causes a shift in the optimal ratio of adsorption/desorption phase time towards adsorption.

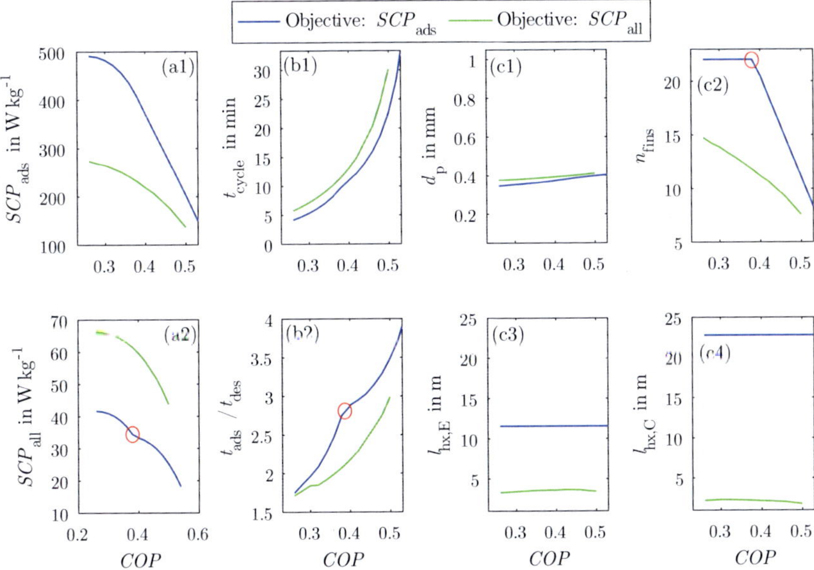

Figure 6.6: Comparison of performance, optimal design, and optimal control for optimisations regarding two different objective functions, SCP_{ads} (blue) and SCP_{all} (green). (left: Objectives) Performance measured in SCP_{ads} (a1), performance measured in SCP_{all} (a2); (middle: Control) Optimal overall cycle time (b1), optimal t_{ads}/t_{des} ratio (b2); (right: Design) optimal particle diameter (c1), optimal fin number (c2), optimal evaporator length (c3), and optimal condenser length (c4).

The section illustrates that the optimal component sizing and adsorber design strongly depends on the objective function. Therefore, it is crucial to define an objective function which incorporates all costs. For the comprehensive design of an adsorption chiller, SCP_{ads} is not suitable as objective function. In the following parts of this thesis, only SCP_{all} is used for optimisation.

6.3 Performance of optimised design with respect to component sizing, fin number, and grain size

The aim of this section is to separate the effects of adsorber-bed design and component sizing. For this purpose, first, an optimisation is conducted only including the adsorber-bed design (grain size and fin number), while keeping the evaporator and condenser heat exchanger length constant. In a second step, also the size of evaporator and condenser are added to the optimisation problem.

Figure 6.7 shows the Pareto frontiers regarding SCP_{all} and the associated optimal control and design parameters for the two optimisation steps.

By optimising only the adsorber heat exchanger design (red lines), the maximum achievable SCP_{all} rises by 24 % to $SCP_{\text{all,max}} = 62.6\,\text{W}\,\text{kg}^{-1}$ compared to the reference design. In terms of optimised parameters, the optimal cycle times are shorter compared to the reference design, but the adsorption/desorption ratio increases for high specific cooling powers and low COPs. The fin number starts at $n_{\text{fins}} = 15$ for the highest SCP_{all} and drops to 6 to achieve the highest COP possible.

By including also sizing into the optimisation problem, an additional increase of 6 % in maximum SCP_{all} can be achieved by increasing the evaporator size and reducing the condenser size. The optimal overall cycle time further drops because of the lower resistances in the evaporator. The ratio of adsorption/desorption phase times also drops due to the lower resistances in the evaporator. The fin number in the adsorber heat exchanger has the same trend: smaller fin number for higher COPs, but the optimal the optimal fin number is increased by 1 fin compared to the prior optimisation step.

The analysis shows that the optimal adsorber design depends on sizing of evaporator and condenser, although the general trends are similar. It is also evident that a comprehensive optimisation is necessary to achieve the best adsorption chiller performance possible.

Finally, a detailed comparison of the $SCP_{\text{all,max}}$ solutions of initial and fully optimised design is conducted. Figure 6.8 shows the achieved heat flow rates at evaporator and condenser for both designs. Also, the difference in adsorption potential is shown, allowing to see improvements and to identify further bottlenecks. Compared to the reference design (cf. Figure 6.5), the shares of the resistances are more balanced. Still 2 resistances are dominant during the adsorption phase: (1) heat transfer from adsorber heat exchanger to adsorbent and (2) heat transfer from evaporator tubes to working fluid. The heat transfer from adsorber heat exchanger to adsorbent could be increased

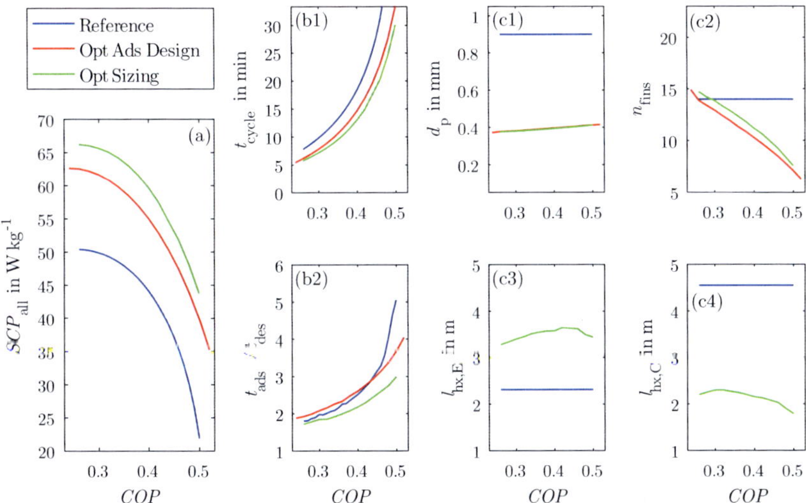

Figure 6.7: Comparison of performance for a successively conducted optimisation. First, the adsorber-bed design is solely optimised (red) and compared to the initial design (blue). Afterwards, also the evaporator and condenser heat exchanger size is included into optimisation (green). (left: objectives) Performance measured in SCP_{all} (a); (middle: control) Optimal overall cycle time (b1), optimal $t_{\text{ads}}/t_{\text{des}}$ ratio (b2); (right: design) optimal particle diameter (c1), optimal fin number (c2), optimal evaporator length (c3), and optimal condenser length (c4).

by additional fins. Since the optimisation did not go to a higher fin number, the cost for additional fins regarding SCP_{all} are too high. Alternatively, the heat transfer from evaporator tubes to working fluid could be enhanced by different evaporator concepts, e. g. coated evaporator tubes as reported by Lanzerath et al. (2016). The condenser itself is well balanced (share of 3 red areas during desorption).

In this chapter, a simultaneous multi-objective optimisation of adsorber-bed design, evaporator size and condenser size has been conducted and further optimisation potential has been identified. Until now, working pair and cycle design are not considered. In the next chapter, a second working pair is added to the analysis answering the question: How much do the equilibria data of the working pair influence optimal adsorber-bed design and component sizing?

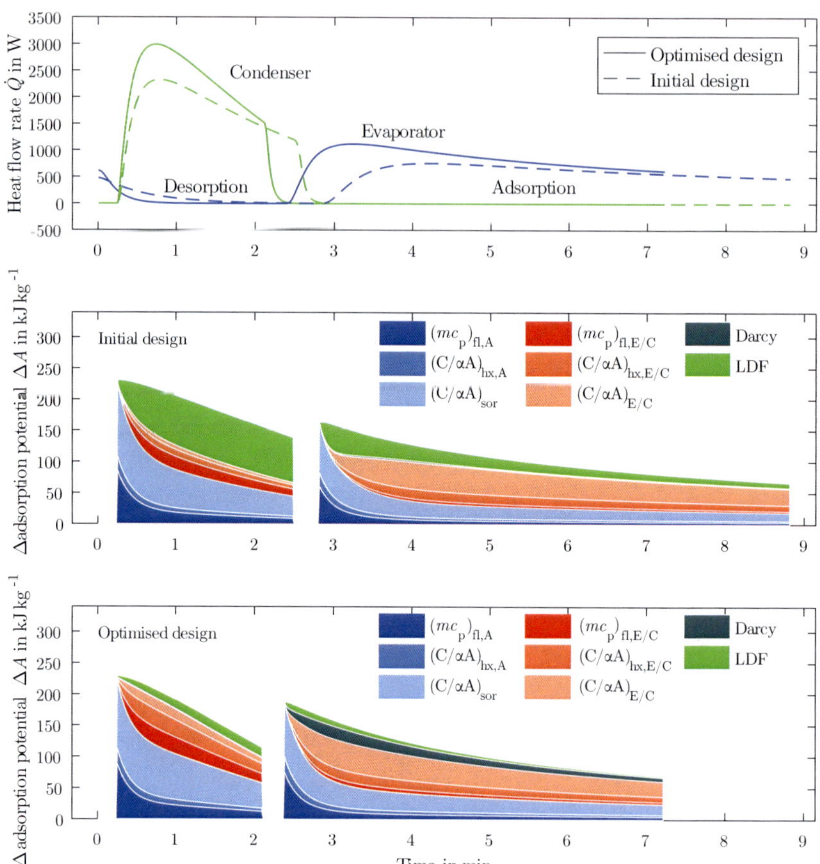

Figure 6.8: Comparison of initial ($d_p = 0.9$ mm, $n_{fins} = 14$, $l_{hx,E} = 2.31$ m, $l_{hx,C} = 4.55$ m) and optimised design ($d_p = 0.375$ mm, $n_{fins} = 14.7$, $l_{hx,E} = 3.27$ m, $l_{hx,C} = 2.20$ m) at $SCP_{all,max}$ operation (cf. Figure 6.7). (top) Heat flow rates of evaporator and condenser. (middle) Initial design: loss of adsorption potential ΔA due to heat and mass transfer resistances (cf. Figure 5.12). (bottom) Optimised design: loss of adsorption potential ΔA due to heat and mass transfer resistances.

CHAPTER 7

The impact of working pair on optimal design and control

In the last two chapters, silica gel 123 / water has been used as working pair. As described in Section 2.2.1, the working pair has a great effect on performance due to its equilibrium data and its kinetics. In this section, the effects of changed equilibria data on system performance and on optimal design and sizing are investigated. For this purpose, a second working pair is added to the analysis, AQSOA Z02 / water.

AQSOA Z02 has been developed by Mitsubishi Plastic Incorporation for desiccant and adsorption chiller and heat pump systems. AQSOA Z02 is a silico aluminophosphate zeolite (Goldsworthy, 2014). AQSOA Z02 has been experimentally tested by Sapienza et al. (2016) in a three-bed adsorption chiller with promising results.

Figure 6.1 shows the discrete design choices and optimisation parameters considered in this chapter.

Due to the object-oriented structure of the Adsorption Energy Systems Library, the working pair can be exchanged easily. The model presented in Figure 6.2 is not changed, only the parameters of the characteristic curve are altered (Section 3.1). The implemented characteristic curve is based on measurements by Goldsworthy (2014). It has to be noted that in this analysis, only the equilibria data of the working pair are changed. Additional effects of the working pair on heat and mass transfer coefficients are not considered. Thus, the influence of equilibria data can be solely investigated.

The analysis is conducted in two steps: First, both working pairs are compared at the temperature triple $T_{\text{high}} = 90\,°\text{C}$, $T_{\text{mid}} = 30\,°\text{C}$, and $T_{\text{low}} = 10\,°\text{C}$. For this temperatures, the maximum loading difference and the shape of the characteristic curves vary. To exclude the effect of maximum loading difference, a second analysis is conducted at a temperature triple of $T_{\text{high}} = 90\,°\text{C}$, $T_{\text{mid}} = 30\,°\text{C}$, and $T_{\text{low}} = 16\,°\text{C}$. For these temperatures, the maximum loading difference for both materials is almost

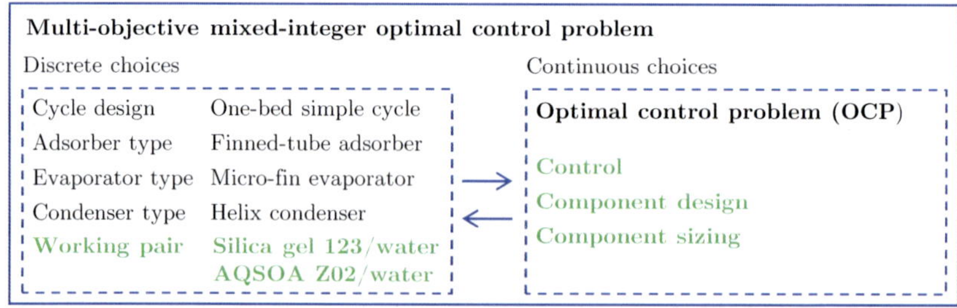

Figure 7.1: Discrete choices and considered optimisation parameters (green) used for analysing the effect of working pair on adsorber-bed design, component sizing, and control.: optimisation of finned-tube adsorber, micro-fin evaporator, helix condenser, and one-bed simple cycle for silica gel 123 / water and AQSOA Z02 / water.

equal, but the shape of the characteristic curve is different.

7.1 Comparing working pairs at initial temperature set: 90-30-10

Figure 7.2 shows the equilibria data of both working pairs. Additionally, the process conditions for the temperature sets (90-30-10 and 90-30-16) and the maximum loading differences are illustrated. For the temperature set 90-30-10, which is also used in Chapter 5 and Chapter 6, it can be observed that AQSOA Z02 has a higher maximum filled pore volume difference between adsorption and desorption compared to silica gel 123 ($\Delta W_{\text{max,Z02}} = 0.218$ and $\Delta W_{\text{max,SG123}} = 0.138$). Therefore, a higher ratio between latent enthalpy and sensible heat can be achieved, leading to a higher maximum COP.

In addition to the difference in maximum filled pore volume difference, the shape of the characteristic curves differs between silica gel 123 and AQSOA Z02. The characteristic curve of silica gel 123 has a convex shape within the boundaries of the temperature set. The characteristic curve of AQSOA Z02 has a S-shape within the boundaries. The step of the "S", i.e. the change in loading, is mainly occurring between adsorption potentials A of $250\,\text{kJ}\,\text{kg}^{-1}$ and $400\,\text{kJ}\,\text{kg}^{-1}$. The step lies perfectly between the minimum and maximum adsorption potential of the application ($A = 170\,\text{kJ}\,\text{kg}^{-1}$ to $470\,\text{kJ}\,\text{kg}^{-1}$).

This differences in equilibrium data have some implications on control and design,

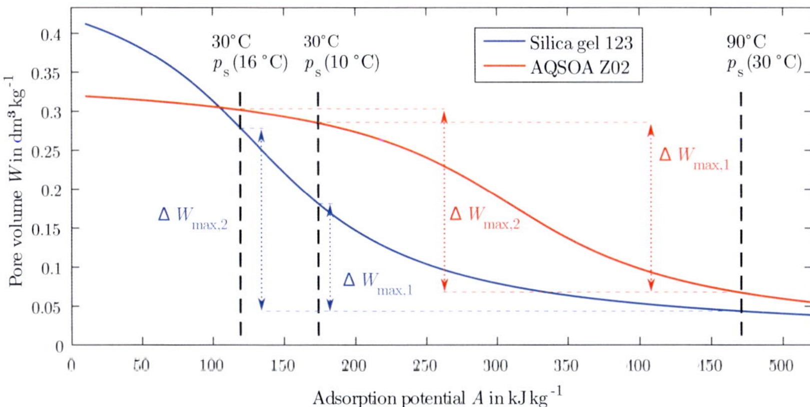

Figure 7.2: Comparison of equilibria data of silica gel 123 (blue) and AQSOA Z02 (red) using the characteristic curve. In addition to the characteristic curve, the process boundaries of two temperature sets, (1) 90-30-10 and (2) 90-30-16, are shown. The maximum difference in loading (filled pore volume) for each working pair is indicated by arrows. For temperature set 1: $\Delta W_{\text{max,SG123,1}} = 0.138\,\text{dm}\,\text{kg}^{-3}$ and $\Delta W_{\text{max,Z02,1}} = 0.218\,\text{dm}\,\text{kg}^{-3}$, for temperature set 2: $\Delta W_{\text{max,SG123,2}} = 0.235\,\text{dm}\,\text{kg}^{-3}$ and $\Delta W_{\text{max,Z02,2}} = 0.234\,\text{dm}\,\text{kg}^{-3}$.

which are discussed in the following:

Regarding control, a different ratio of adsorption/desorption phase times is expected. As shown in Figure 5.11, the convex shape of the characteristic curve leads to a lower average adsorption potential difference for adsorption compared to desorption. Thus, the optimal adsorption and desorption phase time ratio is expected to be $t_{\text{ads}}/t_{\text{des}} > 1$. The S-shape leads to a better balanced adsorption potential for adsorption and desorption compared to the convex shape. Thus, the S-shaped characteristic curve implies more equal times for adsorption and desorption.

To anticipate the effects on design, it is important to identify the new bottlenecks which are caused by the changes in equilibrium data: (1) The higher possible loading uptake of AQSOA Z02 compared to silica gel 123, as described above, leads to higher heat flow rates during adsorption and desorption. To provide or remove the additional heat during adsorption, desorption, vaporisation, and condensation, larger heat exchangers in evaporator and condenser and a higher fin number in the adsorber bed are expected. (2) The higher possible loading uptake of AQSOA Z02 compared to silica gel 123 also leads to a higher possible flow rate of the working fluid in the adsorber bed. This would increase the inter-particle mass-transfer resistance. A larger grain size would counteract this effect. (3) Finally, the S-shape of the characteristic curve of AQSOA Z02, which leads to a better balanced adsorption potential, could shift the heat exchanger sizing: Compared to the case of silica gel, the optimal condenser may be larger and the optimal evaporator may be smaller due to the shift in adsorption potential.

To quantify the expected effects on control and design, we conduct a multi-objective optimisation for AQSOA Z02, optimising the same control and design parameters as in Section 6 and compare them to the results for silica gel 123.

Figure 7.3 shows the results for the multi-objective optimisation: AQSOA Z02 outperforms silica gel 123 in both COP and SCP_{all}. The maximum COP rises from $COP_{\text{max,SG123}} = 0.5$ to $COP_{\text{max,Z02}} = 0.61$. The maximum SCP_{all} rises from $SCP_{\text{all,max,SG123}} = 66.2\,\text{W}\,\text{kg}^{-1}$ to $SCP_{\text{all,max,Z02}} = 80.2\,\text{W}\,\text{kg}^{-1}$.

Regarding control, the ratio of adsorption to desorption phase times is around $t_{\text{ads}}/t_{\text{des}} \approx 1.5$ for AQSOA Z02, whereas it is between 1.7 and 3 for silica gel 123 (Figure 7.3 (b2)). This can be explained by the larger amount of uptake, resulting in a higher ratio of useful latent enthalpy to unused sensible heat (cf. Equation (2.16)).

For the particle diameter, the expected rise from $d_{\text{p}} \approx 0.4$ for silica gel 123 to $d_{\text{p}} \approx 0.5$ for AQSOA Z02 can be observed (Figure 7.3 (c1)).

The fin number shows for both working pairs the same trend: less fin numbers for higher COP values. Nevertheless, for AQSOA Z02, the optimal fin number for the maximum SCP_{all} is $n_{\text{fins}} = 16.25$, which is higher than for silica gel 123 $n_{\text{fins}} = 14.75$ (Figure 7.3 (c2)).

For the heat exchanger size of evaporator and condenser, it is noticeable that the optimal condenser heat exchanger is 50 % larger than for silica gel 123 (Figure 7.3 (c4)). The optimal size of the evaporator rises only slightly (Figure 7.3 (c3)), which is probably caused by the two concurring effects: shifted adsorption potential due to S-shape favouring a smaller evaporator and higher amount of vaporised working fluid favouring a larger evaporator.

As already noticed in Figure 6.4, also in Figure 7.3, a scattering in the optimal parameters can be observed. This scattering is caused by parameters compensating each other: different sets of parameters may lead to similar specific cooling powers within the allowed tolerance. Thus, when evaluating the Pareto frontier by multiple optimisations, the observed scattering occurs. Since the scattering is in a small range, it does not affect the findings discussed above.

The comparison shows that it is necessary to re-optimise control and design for a changing working pair, since all parameters are directly affected by the working pair.

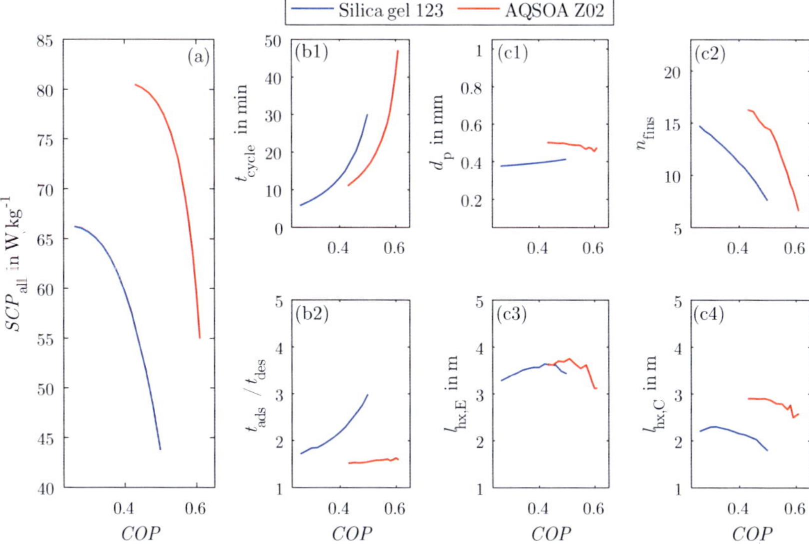

Figure 7.3: Comparison of performance, optimal design, and optimal control for two adsorbents: silica gel 123 (blue) and AQSOA Z02 (red). (left: objectives) Performance measured in SCP_{all} (a); (middle: control) Optimal overall cycle time (b1), optimal t_{ads}/t_{des} ratio (b2); (right: design) optimal particle diameter (c1), optimal fin number (c2), optimal evaporator length (c3), and optimal condenser length (c4).

7.2 The influence of the characteristic curve's shape on performance, design, and control

In Section 7.1, the two working pairs silica gel 123 and AQSOA Z02 were investigated for the temperature set 90-30-10. For this temperature set, the maximum loading difference of AQSOA Z02 is 57 % higher than for silica gel 123. Thus, a better performance of AQSOA Z02 could be expected. In this section, both working pairs are investigated for the temperature set 90-30-16, because for these temperatures the working pairs have an equal maximum loading difference. For this case, the only difference left between the two compared systems is the shape of the characteristic curve (cf. Figure 7.2).

Figure 7.4 shows the optimisation results for both working pairs for the temperature set 90-30-16. For comparison, also the results for 90-30-10 are shown.

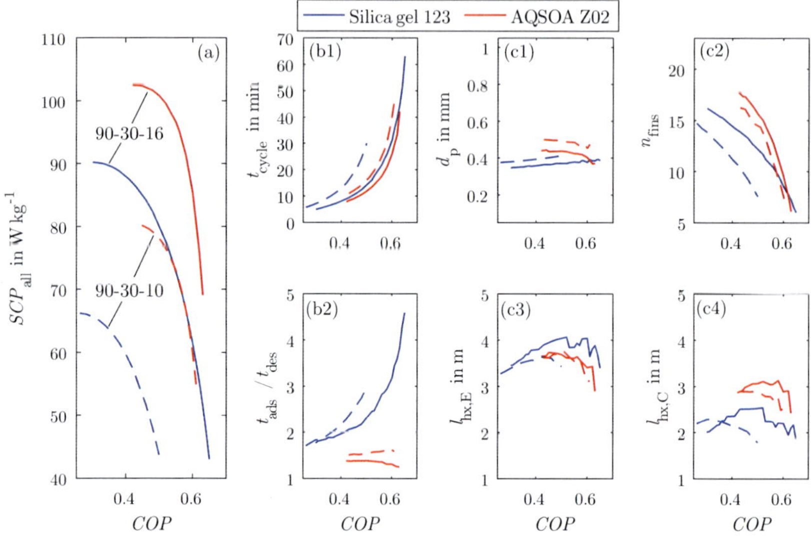

Figure 7.4: Comparison of performance, optimal design, and optimal control for two adsorbents: silica gel 123 (blue) and AQSOA Z02 (red) and for two temperature sets: 90-30-10 (dashed lines) and 90-30-16 (solid lines). (left: Objectives) Performance measured in SCP_{all} (a); (middle: Control) Optimal overall cycle time (b1), optimal $t_{\mathrm{ads}}/t_{\mathrm{des}}$ ratio (b2); (right: Design) optimal particle diameter (c1), optimal fin number (c2), optimal evaporator length (c3), and optimal condenser length (c4).

Both working pairs reach the same maximum $COP_{\mathrm{max}} = 0.65$, which can be explained by the equal maximum loading difference. For both working pairs, this maximum COP is achieved with long cycle times and the minimum fin number $n_{\mathrm{fins}} = 6$.

Regarding SCP_{all}, AQSOA Z02 still outperforms silica gel 123. The reason for the higher specific cooling power is the shape of the characteristic curve. The S-shape leads to more equal adsorption/desorption phase times (Figure 7.4 (b2)), which means there is no bottleneck in adsorption or desorption.

Concerning optimal design, changes for all design parameters can be observed, but with different significance:

The optimal particle diameter is reduced for both, silica gel 123 and AQSOA Z02 (Figure 7.4 (c1)). This reduction is caused by the higher pressure in the evaporator $p_s(16\,^\circ\mathrm{C})$, which causes a higher density of working fluid during adsorption. The higher density leads to a lower velocity of the working fluid in the packed bed. Thus, the inter-particle pressure loss is reduced and the optimal solution shifts to smaller grain sizes.

The optimal fin number increases for both silica gel 123 and AQSOA Z02 (Figure 7.4 (c2)). The effect for silica gel 123 is more pronounced, since the increase of maximum uptake for silica gel 123 is higher than for AQSOA Z02. The increase in uptake leads to higher heat flow rates during adsorption which have to be transferred via the heat exchanger. Thus, a larger heat exchanger area is optimal leading to a higher number of fins.

For evaporator and condenser heat exchanger, the effect is small (Figure 7.4 (c3) and (c4)). Both heat exchangers should be increased slightly to cope with the higher amount of heat transferred during vaporisation and condensation. Again, the effect is more present for silica gel 123 due to the larger increase of maximal loading difference. The jagged lines for high COPs show that the optimal solution is not very sensitive to the heat exchanger length in this range.

This section points out the importance of dynamic considerations to evaluate adsorption chillers: the maximum loading difference is not sufficient to evaluate working pairs. Also, the strong dependence of optimal control and design is illustrated for varying input conditions.

Chapter 8

Two-bed cycle designs at optimal component design and control

In Chapters 5 - 7, rigorous assessment and simultaneous optimisation of design and control has been exemplified for the simple one-bed cycle. For two reasons, in practice, often two-bed adsorption chillers are employed (Wang et al., 2014a): (1) more continuous cooling power because one bed is always in adsorption mode, (2) enhanced performance is possible using sophisticated cycle designs, e. g. heat recovery schemes.

In Section 2.2.3, an overview of possible cycle designs is given. In this chapter, a new cycle design is proposed: a combination of a passive heat recovery scheme with independent phase times of adsorption and desorption. The new cycle design is assessed using the method proposed in Chapter 5 and compared with cycle designs from literature. Figure 8.1 illustrates the discrete design choices and the considered optimisation parameters in this chapter.

Contents of this chapter have been published in:

Bau, U., Gibelhaus, A., Lanzerath, F., and Bardow, A. (2017). Optimal operation of two-bed adsorption chillers by combining heat recovery and reallocation of adsorption/desorption times. In *Proceedings of International Sorption Heat Pump Conference 2017*.

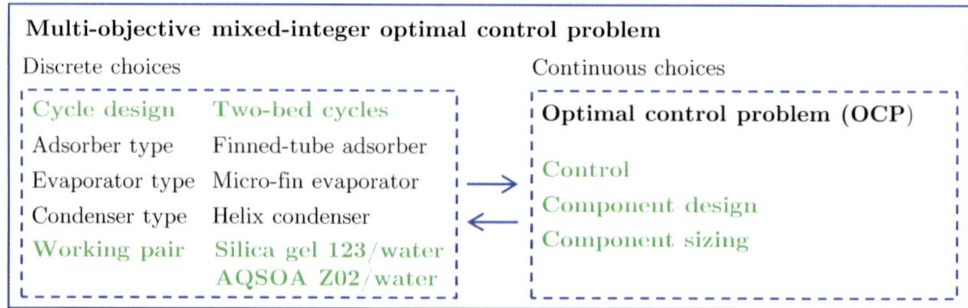

Figure 8.1: Discrete choices and considered optimisation parameters (green) used for comparing cycle designs at optimal design, component sizing, and control.

In Section 8.1, the investigated cycle designs are presented, followed by an analysis of maximum achievable *COP* values (Section 8.2). In Section 8.3, the cycles are compared by conducting a dynamic optimisation of control for each cycle, given a fixed adsorber-bed design and a fixed component sizing. Finally, in Section 8.4, a simultaneous optimisation of control, adsorber-bed design and control is conducted to evaluate the performance limits for the considered design options.

8.1 Two-bed adsorption-chiller-cycle designs

In this section, the new cycle design is derived as a combination of a passive heat recovery scheme and a reallocation of adsorption and desorption phase times.

In adsorption chillers, heat recovery schemes are employed to internally recover parts of the internal energy and of the enthalpy of adsorption. Thus, heat recovery schemes are used to improve *COP*. In Section 2.2.3, two competing heat recovery schemes are presented: the passive heat recovery scheme and the active heat recovery scheme.

To improve power density, a reallocation of phase times is suggested by Sapienza et al. (2011) and Aristov et al. (2012b). This reallocation of adsorption and desorption phase times increases power density by harmonising adsorption/desorption characteristics and phase times. In Chapters 5 - 7, this reallocation has already been used for the one-bed cycle showing that a phase time ratio $t_{\mathrm{ads}}/t_{\mathrm{des}} \neq 1$ is optimal for the investigated designs.

To improve both efficiency and power density, a combination of heat recovery and independent phase times is suggested. Unfortunately, only the passive heat recovery scheme can be combined with independent phase times. For the active heat recovery scheme, equal phase times are necessary (cf. Section 2.2.3). Figure 8.2 illustrates the heat recovery and the phase time allocation options. The possible combinations lead to 5 cycle designs which are considered in this chapter:

1. No heat recovery with equal phase times (NoHR+EPT)
2. No heat recovery with independent phase times (NoHR+IPT)
3. Active heat recovery with equal phase times (AHR)
4. Passive heat recovery with equal phase times (PHR+EPT)
5. Passive heat recovery with independent phase times (PHR+IPT)

In the following, benefits and potential drawbacks for each cycle design are discussed.

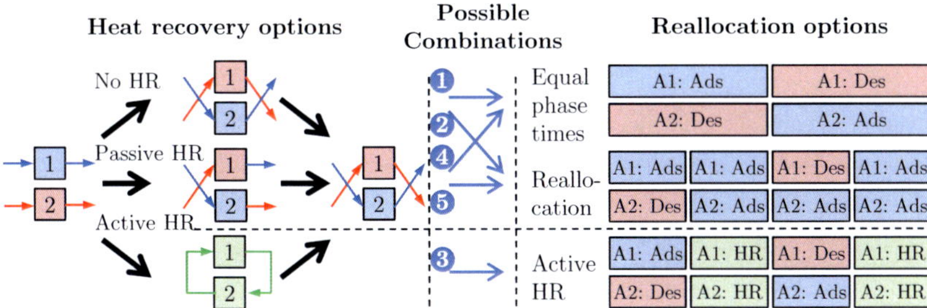

Figure 8.2: Left: Heat Recovery (HR) options (No HR, Passive HR and Active HR). Right: Phase time allocation options (Equal phase times, Reallocation of adsorption and desorption times and Active HR allocation). Middle: Possible combinations of HR scheme and phase time allocation, resulting in 5 operating schemes.

No heat recovery with equal phase times (reference cycle, 1)

Each two-bed cycle studied in this thesis consists of 1 evaporator, 1 condenser, and 2 adsorbers. Each adsorber is connected to the evaporator and condenser by flap valves. Thus, when the pressure in one adsorber is higher than in the condenser, the flap valve opens and working fluid flows to the condenser. Also, when the pressure in the evaporator is higher than in one adsorber, the flap valve opens and the working fluid flows to this adsorber. Condenser and evaporator are connected by a return valve. This configuration is shown in Figure 2.7b. The evaporator is continuously connected to the cooling circuit at T_{low} while the condenser is continuously connected to the heat sink at T_{mid}. The 5 cycle designs only differ in the control of the heat exchanger fluid circuit of the two adsorber beds.

The cycle with no heat recovery and equal phase times consists of two equally long phases which are illustrated in Figure 8.2: In phase 1, Adsorber 1 is connected to the heat sink at T_{mid} and thus, it is in adsorption mode. Adsorber 2 is connected to the heat source at T_{high} and thus, it is in desorption mode. In phase two, adsorber 1 is connected to the heat source and adsorber 2 is connected to the heat sink. Additionally to Figure 8.2, the heat exchanger fluid circuits for all phases are illustrated in Figure C.1.

The main benefit of this cycle is its simplicity. The heat source and heat sink are always connected to one adsorber, which ensures an easy hydraulic circuit. This cycle neither considers the characteristics of adsorption nor those of desorption and it also does not employ any heat recovery scheme.

No heat recovery with independent phase times (2)

For the second cycle, reallocation of phase times is added. To do so, an additional phase is added where both adsorbers are in adsortpion mode (when desorption kinetics are faster) or in desorption mode (when adsorption kinetics are faster). For the case that desorption is faster, all 4 phases of this cycle design are shown in Figure 8.2. Additionally, the heat exchanger fluid circuits for all phases are illustrated in Figure C.2.

The benefit of this cycle is that it fits the specific characteristics of adsorption and desorption kinetics. Thus, a higher SCP_{all} is expected. For the COP, no improvements are expected compared to the cycle with equal phase times. As a downside, the additional challenge for the hydraulic circuit should be mentioned: When both adsorbers are connected to the evaporator, none is connected to the heat source.

Active heat recovery with equal phase times (3)

The third investigated cycle design is the active heat recovery scheme with equal cycle times. The active heat recovery aims to recover the internal energy of adsorbent, heat exchanger metal, heat exchanger fluid, and adsorption enthalpy. To do so, the fluid circuits of the adsorbers are connected when one adsorber has completed adsorption and the other adsorber has completed desorption. The maximum heat is recovered when both adsorbers have the same temperature. After the active heat recovery phase, the adsorbers are, again, connected to the heat sink or the heat source. Figure 8.2 shows the phases of the active heat recovery scheme for a two-bed chiller. Additionally, the heat exchanger fluid circuits for all phases are illustrated in Figure C.3.

The benefit of the active heat recovery scheme is that it may recover large amounts of the energy necessary for desorption. Thus, a significant increase in efficiency is possible. At the same time, the power density is expected to decrease due to the smaller driving forces during heat recovery.

At the end of the heat recovery phase, heat exchanger fluids in both adsorbers have equal temperatures. Thus, the active heat recovery scheme cannot be used in combination with the passive heat recovery scheme (cf. Section 2.2.3). To fully use active heat recovery, the largest temperature gap between adsorbers has to be used. This means that one adsorber has to be at the end of adsorption and the other has to be at the end of desorption. Therefore, the active heat recovery scheme requires equal phase times. Independent phase times could only be achieved with an additional thermal energy storage, which is not considered in this work.

Another downside of the active heat recovery scheme is an increased complexity in the hydraulic circuit. Depending on the realisation, also an additional pump may be necessary. In this work, only the thermodynamic aspects are considered.

Passive heat recover with equal phase times (4)

The fourth investigated cycle design is the passive heat recovery cycle with equal cycle times proposed by Wang et al. (2005b). In this heat recovery scheme, the internal energy of the fluid is recovered, as explained in Section 2.2.3. This can be achieved by an additional phase, when changing from adsorption to desorption mode or from desorption to adsorption mode. In the additional phase, the inlet of the adsorber is already connected to the heat source, whereas the outlet of the adsorber is still connected to the heat sink. By this, the cold fluid in the adsorber is flowing back to the heat sink and not to the heat source, as in the simple cycle. The cold fluid (T_{mid}) has not be heated by the heat source and the internal energy of the fluid is recovered. The times of the additional phases 2 and 4 should be as long as it takes to replace the heat exchanger fluid in the adsorber. They are, therefore, determined by the adsorber heat exchanger length and the volume flow rate. The additional phase is illustrated in Figure 8.2. Additionally, the heat exchanger fluid circuits for all phases are illustrated in Figure C.4.

The passive heat recovery scheme enhances efficiency without loosing power density. Power density is not influenced, since the heat recovery does not take additional time. Merely the valves in the heat exchanger fluid have to be controlled in a smarter way which may lead to a slightly more complex control strategy. As a downside, the heat recovery rate is limited to the internal energy of the heat exchanger fluid. The achievable maximum COP is discussed in the next Section 8.2.

Passive heat recovery with independent phase times (new proposed cycle, 5)

As a new cycle design, the combination of the passive heat recovery scheme with independent phase times is suggested. This scheme combines the benefits of an increased COP and an increased SCP_{all}. To do so, the cycle without heat recovery and with independent phase times is enhanced by 4 additional phases to recover the internal energy of the heat exchanger fluid. The proposed new cycle design is shown in Figure 8.2. Additionally, the heat exchanger fluid circuits for all phases are illustrated in Figure C.5.

To realise this cycle design, additional buffers are necessary in the heat exchanger fluid circuits: when both adsorber inlets are connected to the heat sink, one adsorber outlet is still connected to the heat source to recover the internal energy of the heat exchanger fluid (cf. Figure C.5, Phase 2). Since the heat exchanger fluid is returning to the heat source, and the heat source is not connected to an adsorber inlet, the amount of heat exchanger fluid at the heat source increases. At the same time, the amount of heat exchanger fluid at the heat sink decreases, since both adsorber inlets and only one adsorber outlet are connected to the heat sink. To compensate for these fluctuations in the amount of heat exchanger fluid at heat source and heat sink, two buffers are needed as illustrated in Figure C.5.

In the next section, the maximum achievable COPs are derived for the three heat recovery schemes: no heat recovery, active heat recovery and passive recovery, which is a function of the working pair and the adsorber-bed design. For comparison, also the maximum achievable COP is calculated, which is only dependent on the working pair.

8.2 Maximum COPs of heat recovery schemes

In the last section, three heat recovery schemes are introduced. In this section, the maximum COP for each heat recovery scheme is derived and compared. The allocation of phase times does not influence the maximum COP, since for the maximum COP the system is assumed to reach equilibrium state. As an upper bound for efficiency, Carnot efficiency is used. Carnot efficiency is only a function of process temperatures.

For a thermally driven chiller, Carnot efficiency COP_{C} can be described as a product of the efficiencies of a Carnot cycle η_C and a reversed Carnot cycle $\eta_{\mathrm{C,r}}$:

$$COP_{\mathrm{C}} = \eta_C \eta_{\mathrm{C,r}} = \frac{T_{\mathrm{high}} - T_{\mathrm{mid}}}{T_{\mathrm{high}}} \frac{T_{\mathrm{low}}}{T_{\mathrm{mid}} - T_{\mathrm{low}}} = \frac{1 - {}^{T_{\mathrm{mid}}}/{}_{T_{\mathrm{high}}}}{{}^{T_{\mathrm{mid}}}/{}_{T_{\mathrm{low}}} - 1} \,. \tag{8.1}$$

To calculate the maximum COP, it is assumed that the adsorption cycle reaches equilibrium, i. e. for desorption: $p_{\mathrm{C}} = p_{\mathrm{s}}(T_{\mathrm{mid}})$ and $T_{\mathrm{ad}} = T_{\mathrm{hx,fl}} = T_{\mathrm{hx}} = T_{\mathrm{high}}$ and for adsorption: $p_{\mathrm{E}} = p_{\mathrm{s}}(T_{\mathrm{low}})$ and $T_{\mathrm{ad}} = T_{\mathrm{hx,fl}} = T_{\mathrm{hx}} = T_{\mathrm{mid}}$.

No heat recovery

For the two-bed cycle without heat recovery, the maximum COP can be calculated using Equation (2.16):

$$COP_{\mathrm{NoHR}} = \frac{\int_{w_{\min}}^{w_{\max}} \left[\Delta h_{\mathrm{v}}(T_{\mathrm{E}}) - \left(h^{\mathrm{l}}(T_{\mathrm{C}}) - h^{\mathrm{l}}(T_{\mathrm{E}})\right)\right] \,\mathrm{d}w}{\underbrace{\int_{T_{\mathrm{ads}}}^{T_{\mathrm{des}}} \left(\frac{m_{\mathrm{hx,fl}}}{m_{\mathrm{sor}}} c_{\mathrm{hx,fl}} + \frac{m_{\mathrm{hx}}}{m_{\mathrm{sor}}} c_{\mathrm{hx}} + c_{\mathrm{sor}} + w(T_{\mathrm{ad}}, p_{\mathrm{ad}}) c_{\mathrm{ad}}\right) \mathrm{d}T}_{\text{sensibe heat input}} + \underbrace{\int_{w_{\min}}^{w_{\max}} \Delta h_{\mathrm{ads}}(T_{\mathrm{ad}}, w_{\mathrm{ad}}) \,\mathrm{d}w}_{\text{desorption enthalpy}}} \,. \tag{8.2}$$

The maximum and minimum loading can be calculated using the equilibrium data of the working pair (Appendix 7). The enthalpy of vaporisation is determined by the evaporator temperature $\Delta h_{\mathrm{v}}(T_{\mathrm{E}})$ and can be assumed constant for the ideal cycle.

To illustrate the amount of heat which can be recovered, the differential heat of adsorption and desorption with respect to the temperature can be used (Meunier, 1985). These differential heats include the sensible heat and the enthalpy of adsorption/desorption. In the following, the differential heat of desorption is derived, which can be done analogously for the differential heat of adsorption. The differential loading with respect to the temperature $\frac{\mathrm{d}w}{\mathrm{d}T}$ can be calculated from equilibrium data. Thus, the differential heat of desorption with respect to the temperature reads:

$$\frac{\mathrm{d}Q_{\mathrm{des}}}{\mathrm{d}T} = \frac{m_{\mathrm{hx,fl}}}{m_{\mathrm{sor}}} c_{\mathrm{hx,fl}} + \frac{m_{\mathrm{hx}}}{m_{\mathrm{sor}}} c_{\mathrm{hx}} + c_{\mathrm{sor}} + w(T_{\mathrm{ad}}, p_{\mathrm{ad}}) c_{\mathrm{ad}} + \Delta h_{\mathrm{ads}}(T_{\mathrm{ad}}, w) \frac{\mathrm{d}w}{\mathrm{d}T} \,, \tag{8.3}$$

leading to:

$$COP_{\mathrm{NoHR}} = \frac{\int_{w_{\min}}^{w_{\max}} \left[\Delta h_{\mathrm{v}}(T_{\mathrm{E}}) - \left(h^{\mathrm{l}}(T_{\mathrm{C}}) - h^{\mathrm{l}}(T_{\mathrm{E}})\right)\right] \,\mathrm{d}w}{\int\limits_{T_{\mathrm{ads}}}^{T_{\mathrm{des}}} \frac{\mathrm{d}Q_{\mathrm{des}}}{\mathrm{d}T} \,\mathrm{d}T} . \tag{8.4}$$

Active heat recovery (AHR)

The active heat recovery scheme (AHR), as described in Section 2.2.3, transfers the energy of the hot adsorber to the cold adsorber for desorption. This heat transfer is possible until both adsorbers have the same temperature T_{AHR}, leading to a:

$$COP_{\mathrm{AHR}} = \frac{\int_{w_{\min}}^{w_{\max}} \left[\Delta h_{\mathrm{v}}(T_{\mathrm{E}}) - \left(h^{\mathrm{l}}(T_{\mathrm{C}}) - h^{\mathrm{l}}(T_{\mathrm{E}})\right)\right] \,\mathrm{d}w}{\int\limits_{T_{\mathrm{AHR}}}^{T_{\mathrm{des}}} \frac{\mathrm{d}Q_{\mathrm{des}}}{\mathrm{d}T} \,\mathrm{d}T} . \tag{8.5}$$

Passive heat recovery (PHR)

The passive heat recovery scheme (PHR), as described in Section 2.2.3, recovers the internal energy of the heat exchanger fluid. Thus, the internal energy of the heat exchanger fluid represented by the heat capacity $C_{\mathrm{hx,fl}}$ is not considered for the COP:

$$COP_{\mathrm{PHR}} = \frac{\int_{w_{\min}}^{w_{\max}} \left[\Delta h_{\mathrm{v}}(T_{\mathrm{E}}) - \left(h^{\mathrm{l}}(T_{\mathrm{C}}) - h^{\mathrm{l}}(T_{\mathrm{E}})\right)\right] \,\mathrm{d}w}{\int\limits_{T_{\mathrm{ads}}}^{T_{\mathrm{des}}} \left(\frac{m_{\mathrm{hx}}}{m_{\mathrm{sor}}} c_{\mathrm{hx}} + c_{\mathrm{sor}} + w(T_{\mathrm{ad}}, p_{\mathrm{ad}}) c_{\mathrm{ad}} + \Delta h_{\mathrm{ads}}(T_{\mathrm{ad}}, w) \frac{\mathrm{d}w}{\mathrm{d}T}\right) \,\mathrm{d}T} . \tag{8.6}$$

Maximum heat recovery ($\mathbf{HR_{max}}$)

The maximum heat recovery ($\mathrm{HR_{max}}$) is achieved, when all the released heat during adsorption is recovered at its specific temperature level. Assuming constant heat capacities, the internal energies can be fully recovered. The recoverable share of the enthalpy of adsorption/desorption depends on the temperature at which this enthalpy is released or needed. The working pair's equilibrium data determines the temperature at which the heat during adsorption is released as well as the temperature at which the heat for desorption is needed. Thus, the maximum achievable COP is only defined by the working pair.

The maximum heat recovery is a theoretical value, since it requires to transfer the heat without a temperature gradient. This results in very low heat flow rates ($\dot{Q} \to 0$)

or very large heat transfer areas ($A \to \infty$).

Figure 8.3 shows the differential heat curves for adsorption and desorption and illustrates the recovered amount of energy for each heat recovery scheme by the filled area. For the passive heat recovery (Figure 8.3 left), only the internal energy of the heat exchanger fluid is recovered. For the active heat recovery scheme (Figure 8.3 middle), the area below the differential heat curve of desorption can be recovered until the temperature T_{AHR} is reached. For the maximum heat recovery scheme (Figure 8.3 right), the area below the minimum of differential heat of desorption and adsorption can be recovered.

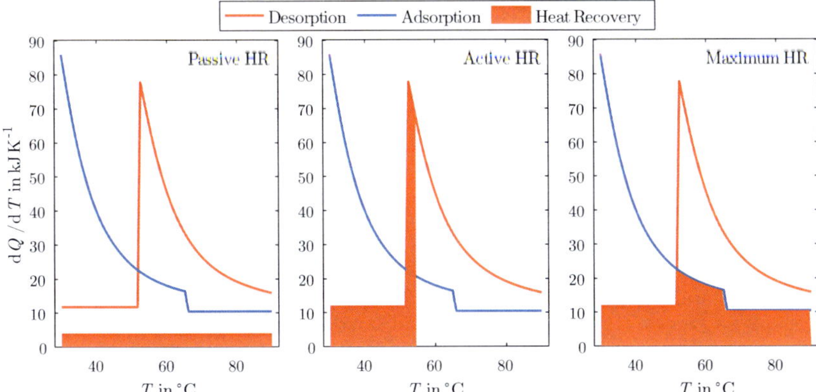

Figure 8.3: Differential heat with respect to temperature of adsorption and desorption. The maximum energy, which can be recovered by each heat recovery scheme, is shown as red area. Left: passive heat recovery scheme; middle: active heat recovery scheme; right: maximum heat recovery scheme.

Achievable COP-values

Figure 8.4 shows the achievable maximum COP values for different heat recovery schemes, adsorber-bed designs, working pairs, and temperature triples. To allow for better comparison between the temperature triples, the COP values are normalised by the achievable Carnot efficiency. It can be observed that the maximum COP level increases depending on the heat recovery scheme: no heat recovery $<$ passive heat recovery $<$ active heat recovery $<$ maximum heat recovery. Even with the maximum heat recovery scheme, only less than 50 % of Carnot efficiency can be achieved.

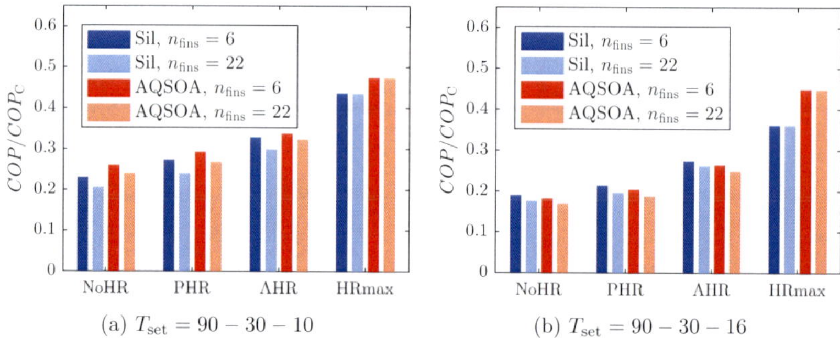

(a) $T_{\text{set}} = 90 - 30 - 10$

(b) $T_{\text{set}} = 90 - 30 - 16$

Figure 8.4: Achievable $^{COP}/_{COP_C}$-value dependent on heat recovery scheme, adsorber design, working pair, and temperature triple. The heat recovery schemes are: no heat recovery (noHR), passive heat recovery (PHR), active heat recovery (AHR), and maximum heat recovery (HR_{max}). For adsorber design, two fin numbers are investigated ($n_{\text{fins}} = 6$ and $n_{\text{fins}} = 22$). As working pairs, silicagel 123 (Sil) and AQSOA Z02 (AQSOA) are used.

Figure 8.4 also illustrates the interdependence between performance and all design levels. For example, the loss of COP when changing the fin number from 6 to 22 depends on the heat recovery scheme. For the temperature set 90-30-10 and silica gel 123, the loss in $^{COP}/_{COP_C}$ for the passive heat recovery scheme is 0.033, for the active heat recovery scheme 0.029 and for the maximum heat recovery scheme it is 0 (since COP_{HRmax} is only a function of working pair).

The quantification of maximum efficiency gives a first indication of preferable designs and it shows the interconnections between the design levels. Still it does not enable the designer to favour a certain design option, since it does not include the effects on power density. Therefore, a comprehensive analysis is conducted in the next sections considering system dynamics and simultaneous optimisation of all design levels.

8.3 Assessment and comparison of cycle designs for a fixed component design and sizing

To investigate the effects of heat recovery schemes on efficiency and power density, a multi-objective optimisation (cf. Section 5.2) is conducted in two steps:

First, in this section, only control of each heat recovery scheme is optimised. The design of adsorber-bed and components is identical for each cycle design. Thus, this section answers the question on how a fixed two-bed adsorption chiller should be operated.

Second, in the next Section 8.4, a simultaneous optimisation of cycle design, adsorber-bed design, and component sizing is conducted. Thus, the next section answers the question: What is the maximum achievable performance within the considered design options?

The necessary dynamic model to conduct the optimisation is shown in Figure 8.5. The model used in Chapters 6 and 7 is extended by a second adsorber. This leads to 16 differential states in total. The energy and mass balance and all algebraic equations are according to the models described in Chapter 3. These model equations are similar for the 5 considered cycle designs. To simulate the cycle designs only the input temperatures of the adsorbers ($T_{\mathrm{fl,A1,in}}$ and $T_{\mathrm{fl,A2,in}}$) differ for the phases. The input temperatures for each phase are listed in Table 8.1. To calculate the COP for the passive heat recovery scheme, the internal energy of the fluid in the heat exchanger fluid is not considered, since it is recovered. Thus, perfect recovery is assumed for both heat recovery schemes.

Table 8.1: Input conditions for adsorber temperatures depending on phase time allocation scheme according to Figure 8.2.

	Phase 1	**Phase 2**	**Phase 3**	**Phase 4**
Equal phase times	$T_{\mathrm{fl,A1,in}}=T_{\mathrm{mid}}$ $T_{\mathrm{fl,A2,in}}=T_{\mathrm{high}}$		$T_{\mathrm{fl,A1,in}}=T_{\mathrm{high}}$ $T_{\mathrm{fl,A2,in}}=T_{\mathrm{mid}}$	
Reallocation (Unequal phase times)	$T_{\mathrm{fl,A1,in}}=T_{\mathrm{mid}}$ $T_{\mathrm{fl,A2,in}}=T_{\mathrm{high}}$	$T_{\mathrm{fl,A1,in}}=T_{\mathrm{mid}}$ $T_{\mathrm{fl,A2,in}}=T_{\mathrm{mid}}$	$T_{\mathrm{fl,A1,in}}=T_{\mathrm{high}}$ $T_{\mathrm{fl,A2,in}}=T_{\mathrm{mid}}$	$T_{\mathrm{fl,A1,in}}=T_{\mathrm{mid}}$ $T_{\mathrm{fl,A2,in}}=T_{\mathrm{mid}}$
Active HR	$T_{\mathrm{fl,A1,in}}=T_{\mathrm{mid}}$ $T_{\mathrm{fl,A2,in}}=T_{\mathrm{high}}$	$T_{\mathrm{fl,A1,in}}=T_{\mathrm{fl,A2,out}}$ $T_{\mathrm{fl,A2,in}}=T_{\mathrm{fl,A1,out}}$	$T_{\mathrm{fl,A1,in}}=T_{\mathrm{high}}$ $T_{\mathrm{fl,A2,in}}=T_{\mathrm{mid}}$	$T_{\mathrm{fl,A1,in}}=T_{\mathrm{fl,A2,out}}$ $T_{\mathrm{fl,A2,in}}=T_{\mathrm{fl,A1,in}}$

To compare the 5 cycle designs for a given fixed component design and sizing, a set of design parameters is derived from Section 6.3. For the adsorber-bed design, a fin number of $n_{\mathrm{fins}} = 15$ and a particle diameter of $d_{\mathrm{p}} = 0.4\,\mathrm{mm}$ is chosen, since this design achieved the highest power density (cf. Figure 6.7). For the evaporator and condenser, twice the size of the $SCP_{\mathrm{all,max}}$ solution is chosen: $l_{\mathrm{hx,E}} = 6.6\,\mathrm{m}$ and $l_{\mathrm{hx,C}} = 4.6\,\mathrm{m}$. Heat and mass transfer coefficients are chosen according to Tables 5.2 and 5.3.

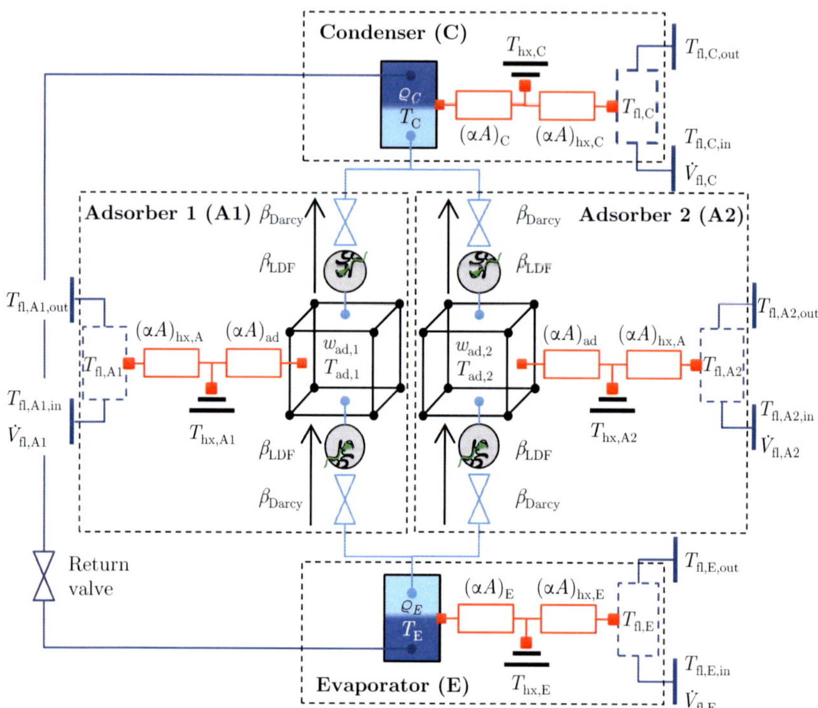

Figure 8.5: Structure of dynamic two-bed chiller model consisting of an evaporator (E), a condenser (C), and two adsorbers (A1 and A2). All differential states (16), and heat and mass transfer coefficients are named.

Figure 8.6 shows the obtained Pareto frontiers of the 5 cycle designs for optimal control. The reference cycle (no heat recovery and equal phase times) has the worst performance. Employing the passive heat recovery scheme, the Pareto frontier shifts to higher *COP* with an increase of 0.08. Reallocation of phase times can enhance the power density by 4 % for both NoHR and PHR. The active heat recovery scheme achieves the same maximum SCP_{all} as the reference cycle. By employing AHR, the cycle outperforms all other cycles for high *COP*s.

It can be observed that there is no dominant cycle which outperforms the others for all operating conditions: For high power densities, the passive heat recovery scheme with independent cycle times should be used. When surpassing $COP = 0.46$, active heat recovery with equal cycle times becomes the favourable cycle. Thus, the operating scheme in the two-bed chiller should be adjusted depending on the cooling load to obtain the best achievable performance. The overall Pareto frontier is marked as a bold line in Figure 8.6.

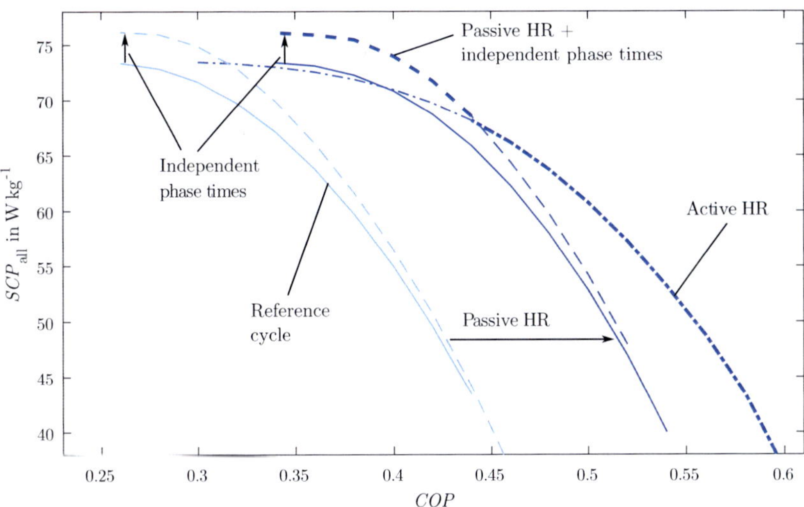

Figure 8.6: Pareto frontier of cycle design for a fixed working pair (silica gel 123), adsorber design ($n_{\text{fins}} = 15$ and $d_{\text{p}} = 0.4\,\text{mm}$), and component sizing ($l_{\text{hx,E}} = 6.6\,\text{m}$ and $l_{\text{hx,C}} = 4.6\,\text{m}$).

8.4 Simultaneous optimisation of cycle designs, adsorber-bed design and component sizing

In the last section, all cycle designs are compared with a fixed component design and sizing. To determine the maximum performance achievable with the considered design options, a simultaneous optimisation of all design parameters (fins, grain size, heat exchanger length) is conducted in this section (cf. Chapters 6 and 7).

Figure 8.7 shows the obtained overall Pareto frontiers. For comparison, also the performance of the initial design is plotted, which has been experimentally validated by Lanzerath (2014). For all three Pareto frontiers, the dominant cycle design switches from passive heat recovery with independent cycle times to active heat recovery when moving to higher *COP*s.

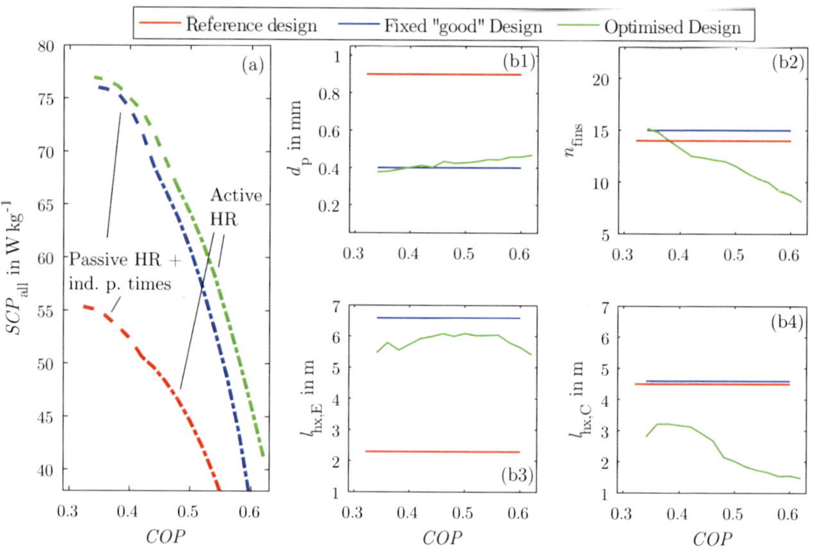

Figure 8.7: Overall Pareto frontier of cycle designs for three cases: reference design ($n_{\text{fins}} = 14$, $d_{\text{p}} = 0.9\,\text{mm}$, $l_{\text{hx,E}} = 2.3\,\text{m}$, and $l_{\text{hx,C}} = 4.5\,\text{m}$), fixed “good” design ($n_{\text{fins}} = 15$, $d_{\text{p}} = 0.4\,\text{mm}$, $l_{\text{hx,E}} = 6.6\,\text{m}$, and $l_{\text{hx,C}} = 4.6\,\text{m}$), and optimised design.

The simultaneous optimisation reveals a maximum $SCP_{\text{all}} = 77\,\text{W}\,\text{kg}^{-1}$ which is only 1 % higher compared to the fixed “good” design investigated in Section 8.3. Thus, the fixed design is almost SCP_{all} optimal, only the condenser has been chosen too large. For higher COP values, the gap in SCP_{all} increases between the fixed and the optimised design. For these operating points, adsorber-bed designs with lower fin numbers are optimal, as already shown in Chapter 5.

Compared to the reference design used by Lanzerath (2014), the $SCP_{\text{all,max}}$ can be increased by 40 % from $SCP_{\text{all}} = 55\,\text{W}\,\text{kg}^{-1}$ to $SCP_{\text{all}} = 77\,\text{W}\,\text{kg}^{-1}$ by decreasing the grain size, increasing the evaporator and decreasing the condenser heat exchanger. The efficiency can be increased from $COP = 0.54$ to $COP = 0.62$ for a comparable $SCP_{\text{all}} \approx 40\,\text{W}\,\text{kg}^{-1}$. These significant increases in both efficiency and power density underline the importance of a rigorous and comprehensive optimisation of all design levels and control.

Finally, the obtained optimisation results for design and control are embedded in the context which is set by the full-scale experiments. Figure 8.8 shows all experimental results with more than 2 data points for a specific set of input conditions (cf. Figure 2.12). Additionally, Figure 8.8 illustrates the Pareto frontier for the initial design with a one-bed cycle design (cf. Figure 6.4) and the overall Pareto frontier of the fixed "good" design with the best two-bed cycle designs. As objective function for the specific cooling power, SCP_{ads} is chosen since SCP_{all} is seldom reported in literature. For this reason, also, fixed "good" design is shown, since it is not effected by the objective function in contrast to the fully optimised design (cf. Section 6.2). It also has to be mentioned that the temperature levels for the reported designs are different which makes a direct comparison of designs difficult. Regardless of these restrictions, Figure 8.8 nicely illustrates the potential of a multi-objective optimisation of design and control.

Chapters 5 - 8 show how important it is to conduct a comprehensive multi-objective analysis. The analysis investigated the effect of all design levels (cf. Figure 2.3) from working pair to cycle design and showed the importance of optimal control. Based on an initial design, the chapters assess and compare designs regarding efficiency and power density, point out the importance of the objective function on design, and show a method to visualise heat and mass transfer resistances using the loss in adsorption potential. Thus, Part II of this thesis outlines the way to a simultaneous optimisation of adsorption chillers regarding all design levels and control, exploiting the reachable limits in terms of efficiency and power density.

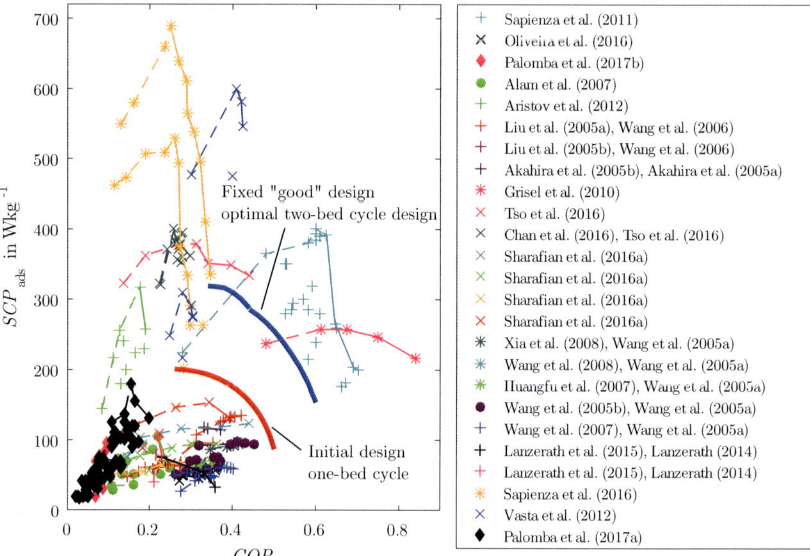

Figure 8.8: Pareto frontiers of reported full-scale experiments with 3 or more data points for a single set of input conditions (temperatures and volume flow rates) and Pareto frontiers of initial design with one-bed cycle (red) and fixed "good" design with optimal two-bed cycle design (blue).

Part III

Control of adsorption energy systems

Chapter 9

Optimal control of adsorption chillers

As shown in Chapters 5 - 8, control of an adsorption chiller strongly influences the system performance measured by power density and efficiency. A control strategy aims at finding the Pareto-optimal phase times depending on the cooling load. As seen in Section 7.2, the optimal control depends on the operating conditions. Therefore, the optimal phase times vary during typical operation scenarios (e. g. fluctuating cooling load, changing solar irradiation, ...).

Control of adsorption chillers is often based on fixed phase times for simplicity reasons (Büttner and Mittelbach, 2013). In this case, however, in most operating points, only sub-optimal phase times are achieved.

As an alternative to fixed cycle times, heuristics and model predictive control are introduced in Section 2.2.4. Heuristics are used to find a good operating point by control criteria based on measured variables, such as evaporator outlet temperature (Büttner and Mittelbach, 2013; Schicktanz, 2011), or heat flow rates (Gräber, 2013).

Contents of this chapter have been published in:

Bau, U., Lanzerath, F., and Bardow, A. (2017). A simple heat-flow-based control strategy for one-bed adsorption chillers leading to Pareto-optimal performance. In *Proceedings of International Sorption Heat Pump Conference 2017*.

A further alternative to fixed phase times is model predictive control (MPC). MPC allows to find the optimal phase times corresponding to a desired objective within a given model accuracy. Compared to heuristics, which are developed for only one cycle design, MPC generally can be applied to any cycle design. These advantages come at the cost of complexity. An MPC for two-bed chillers with equal adsorption and desorption times has been proposed in a previous work by the author (Bau et al., 2015a) and has been evaluated with a simulation study. The simulation study showed that MPC can improve performance for fluctuating input conditions compared to a heuristic-based approach. The simulation study assumed a perfect process model and full knowledge of all system states. Both is difficult to achieve for real world systems.

In this chapter, an MPC for a one-bed chiller is derived and experimentally tested for the first time, including a non-perfect process model and state estimation. Additionally, new and (near) Pareto-optimal heat-flow-based control strategies for one-bed chillers are proposed and experimentally compared to the MPC.

In Section 9.1, the model-predictive-control loop is explained including a test bench, a non-linear process model, a state estimator, and a phase time optimisation problem. In Section 9.2.1, the new heat-flow-based control strategies for one-bed chillers are presented and tested for the chiller model already used in Chapter 6. Finally, in Section 9.3.2, the MPC and the proposed heuristics are experimentally tested for two test cases: a step response and a typical solar cooling case.

9.1 Non-linear model-predictive-control loop

The used model-predictive-control loop consists of the basic elements: plant, process model, state estimator, and optimiser (Rawlings and Mayne, 2009). The loop is shown in Figure 9.1 and explained in the following: The plant (Section 9.1.1), which is a one-bed adsorption chiller in this case, gets external inputs which may fluctuate over time. Control is realised via the phase times for adsorption and desorption. To provide these phase times, a controller is used solving an optimal control problem (OCP). As input for the controller, measurement data (temperatures and volume flows of heat exchanger fluid circuits), a process model (Section 9.1.2), and assumptions of future external inputs are needed. The controller itself (Section 9.1.3) uses a two-step approach: First, all current system states are estimated using historical measurement data. Second, an optimal control problem is solved to find optimal future phase times given the estimated states and the assumed future external inputs. In the following, all parts of the MPC loop are explained in detail.

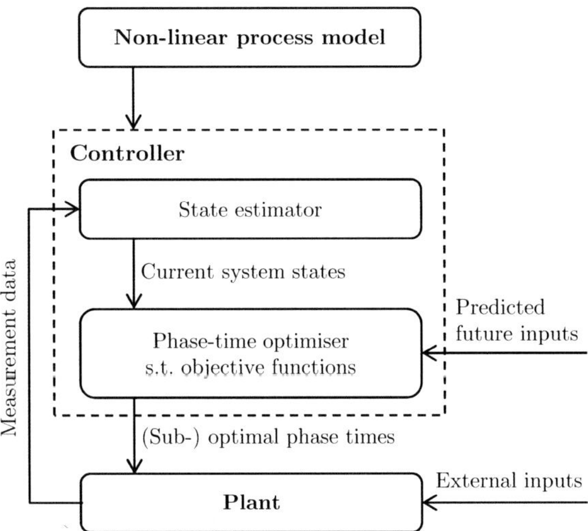

Figure 9.1: Flow sheet of model-predictive-control loop. Based on measurement data from the plant, current system states are estimated. Based on these states and predicted future inputs, phase times are optimised. The control loop is closed by sending the optimal phase times to the plant. Both state estimator and phase-time optimiser use the non-linear process model to predict the system behaviour.

9.1.1 Plant – One-bed adsorption chiller

The model predictive control is experimentally tested for a one-bed adsorption chiller with an evaporator and a condenser. To do so, a modular test bench is employed which has also been used by Lanzerath (2014) to calibrate and validate the model used in Chapter 5 - 8. A scheme of the test bench with all components and sensors is shown in Figure 9.2. The dashed frame indicates the separation between the actual chiller and the test bench used to provide the input temperatures and volume flow rates.

Compared to the components validated by Lanzerath (2014) and used in Chapter 6, the evaporator is replaced by a larger micro-fin evaporator and the condenser is replaced by a compact double-helix heat exchanger. For additional information on the new components see Appendix E.

To determine the coefficient of performance COP and the specific cooling power

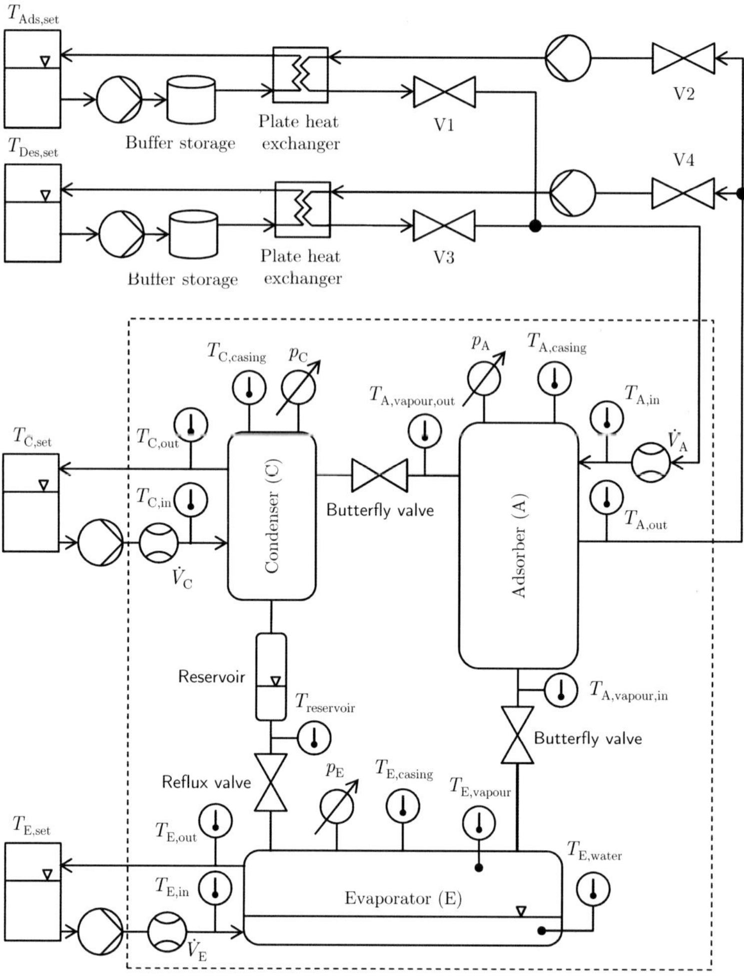

Figure 9.2: Adsorption chiller test bench including thermal baths, pumps, adsorption chiller, and measurement sensors. The dashed line separates adsorption chiller and test bench. For more information on the components, the measurement equipment, and the operation see Appendix E.

SCP_{ads} as described in Section 2.1.3, the cooling energy in the evaporator and the heating input in the adsorber have to be determined. To do so, the enthalpy flow differences in the evaporator and adsorber heat exchangers are employed. The enthalpy flow differences are calculated from the volume flow rates $\dot{V}$ and the temperature differences ΔT:

$$\dot{H}_{i,\text{out}} - \dot{H}_{i,\text{in}} = \dot{V}_i \varrho_i c_i (T_{i,\text{out}} - T_{i,\text{in}}) \tag{9.1}$$

where ϱ is the density and c is the specific heat capacity of the heat exchanger fluid.

By using the enthalpy flow difference of the adsorber heat exchanger as heat input, it is assumed that no heat recovery scheme is employed (cf. Section 8.1). The maximum measurement uncertainty stated as standard deviation for the COP is ± 0.0143 and for the SCP_{ads} it is $\pm 1.69\,\text{W}\,\text{kg}^{-1}$. The measurement uncertainty is calculated by propagation of uncertainty according to the guide to the expression of uncertainty in measurement (Joint Committee for Guides in Metrology, 2008). For further information on the used measurement equipment see also Lanzerath (2014) and Appendix E.

9.1.2 Non-linear process model

For the process model of an MPC, a trade-off exists between model accuracy and calculation effort. Since adsorption chillers are characterised by their intrinsic dynamics, a dynamic non-linear process model is used. The process model is again built with the Adsorption Energy Systems Library in the programming language Modelica (cf. Chapter 3). In principle, the process model is similar to the model described in Chapter 6 with four modifications: (1) heat losses are added, (2) inter-particle mass transfer resistance (Darcy) is neglected because grain sizes of $d_{\text{p}} = 0.9\,\text{mm}$ are used (cf. Section 5.3.3), (3) a pressure loss due to friction in the butterfly valves is added (β_{friction}), and (4) the heat exchanger fluid in the adsorber is modelled with a heat capacity, replacing the liquid cell. This last modification is necessary to incorporate future external inputs as accurately as possible with the given software set-up: process model written in Modelica and optimisation algorithm MUSCOD II written in C++. The final process model is illustrated in (Figure 9.3) including all states and heat and mass transfer coefficients. The model equations of the shown models are according to Chapter 3. For more information on the model changes see Appendix D.

Due to the model changes and the new components, a re-calibration of the model is necessary. Since the phase times provided by an MPC can only be optimal within the accuracy of the process model, it is desirable to have an accurate process model for the whole range of input conditions. Thus, first, the new model is calibrated using

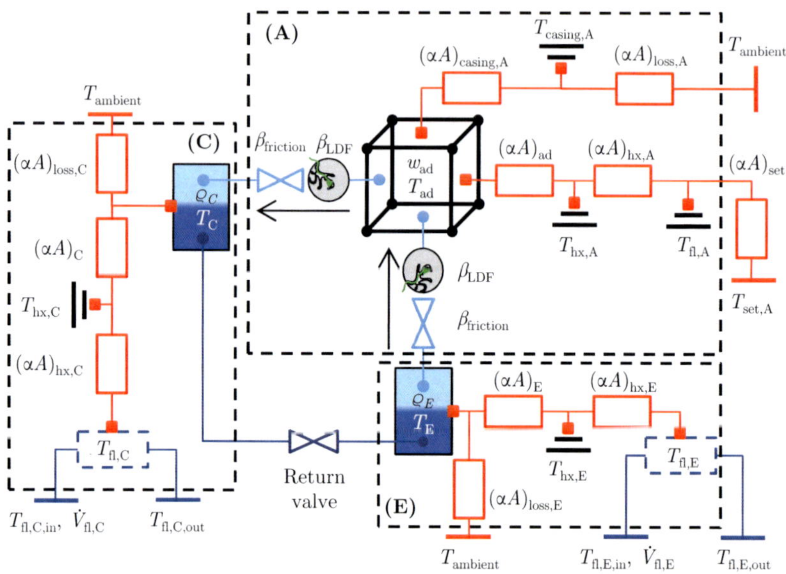

Figure 9.3: Structure of the non-linear process model used for MPC. The process model consists of three components: evaporator (E), condenser (C), and adsorber (A). Heat transfer coefficients and differential states are named.

a calibration cycle. Afterwards, the new model is validated for different phase times and inlet temperatures.

Model calibration is conducted stepwise according to the procedures proposed by Lanzerath (2014) and Schreiber (2017). For more information see again Appendix D.

As criterion for accuracy, the coefficient of variation regarding the measured $\dot{Q}_{\text{meas,i}}$ and simulated heat flow rates $\dot{Q}_{\text{sim,i}}$ is used:

$$CV_{\text{i}} = \frac{\sqrt{\frac{1}{t_{\text{cycle}}} \cdot \int (\dot{Q}_{\text{meas,i}} - \dot{Q}_{\text{sim,i}})^2 \text{d}t}}{\overline{\dot{Q}}_{\text{meas,i}}}. \tag{9.2}$$

For calibration, the following temperatures and cycle times are used: $T_{\text{high}} = 95\,°\text{C}$, $T_{\text{mid}} = 30\,°\text{C}$, $T_{\text{low}} = 10\,°\text{C}$, $t_{\text{ads}} = 450\,\text{s}$, and $t_{\text{des}} = 300\,\text{s}$. The obtained model parameters for the calibration cycle are listed in Table 9.1.

Figure 9.4 shows the achieved accuracy of the simulated evaporator and condenser

Table 9.1: Fitted parameters of the modified non-linear process model used for optimisation.

Parameter		
$\beta_{\mathrm{LDF,ads}}$	5.93e-3	s^{-1}
$\beta_{\mathrm{LDF,des}}$	2.30e-2	s^{-1}
αA_{ad}	$(184 + 2.3 \cdot T_{\mathrm{ad}})$	$\mathrm{W\,K^{-1}}$
αA_{E}	2156	$\mathrm{W\,K^{-1}}$
αA_{C}	2179	$\mathrm{W\,K^{-1}}$
$\alpha A_{\mathrm{casing,A}}$	6	$\mathrm{W\,K^{-1}}$
$\alpha A_{\mathrm{loss,A}}$	1.2	$\mathrm{W\,K^{-1}}$
$\alpha A_{\mathrm{loss,C}}$	1	$\mathrm{W\,K^{-1}}$
$\alpha A_{\mathrm{loss,E}}$	1.5	$\mathrm{W\,K^{-1}}$

heat flow rates for predicted input temperatures. The measured peak in the condenser at the beginning of the desorption phase is underestimated. The main reason for this underestimation lies in the approximation error of the adsorber heat exchanger input temperature. The accuracy is very high for the evaporator heat flow rate, which directly corresponds to the predicted cooling power. The resulting CV_{E} for the calibration is $CV_{\mathrm{E}} = 0.132$. In the next section, the accuracy of the calibrated non-linear process model is validated for varying temperature levels and cycle times.

Validation

To use the calibrated model for control purposes, the model needs to be robust for a wide range of operating conditions (temperatures and cycle times). Therefore, two variations are investigated experimentally: (1) The temperatures levels T_{mid} and T_{high} are varied while keeping the phase times constant. (2) The phase times are varied for a constant temperature set. The resulting accuracy, measured by the CV_{E} is shown in Figure 9.5.

For the temperature variation, the CV_{E} is always below 0.2. For the cycle time variation, the CV_{E} value is growing especially for very long cycles. Since it is very difficult to decide which CV_{E} is acceptable, the deviation of efficiency and power density is also evaluated.

Figure 9.6 shows the measured and simulated COP and SCP_{ads} for varying temperatures and cycle times. For varying temperatures, experiment and simulation show a good agreement with slight deviations for the temperature set (60-25-25-10). For the

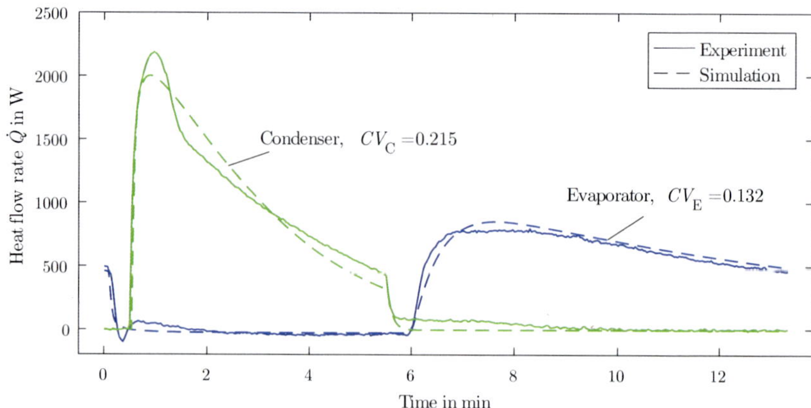

Figure 9.4: Simulation accuracy of the non-linear process model used in the MPC. The simulated (– –) and measured (–) heat flow rates of evaporator and condenser are shown.

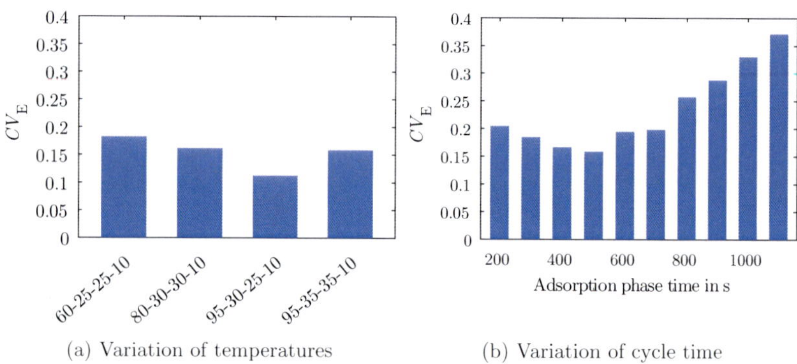

(a) Variation of temperatures

(b) Variation of cycle time

Figure 9.5: CV_E of non-linear process model for different operating conditions. Left: Four temperature combinations for desorption, adsorption, condensation and vaporisation (T_{Des}-T_{Ads}-T_C-T_E in °C). Phase times ($t_{des} = 500\,s$, $t_{ads} = 300\,s$) are kept constant for this variation. Right: Increasing adsorption times from 200 seconds to 1100 seconds. The ratio between adsorption and desorption time is 5/3. The inlet water temperatures are constant (95-35-35-10).

phase time variation, it can be observed that the simulated COP fits almost perfectly for short cycle times and is just above the measurement uncertainty for very long cycle times. The SCP_{ads} is underestimated for short cycle times and slightly overestimated for long cycle times. Still, the trend is the same for simulation and experiment and the maximum SCP_{ads} is observed at similar cycle times. Therefore, the calibrated non-linear process model is regarded suitable for the MPC loop.

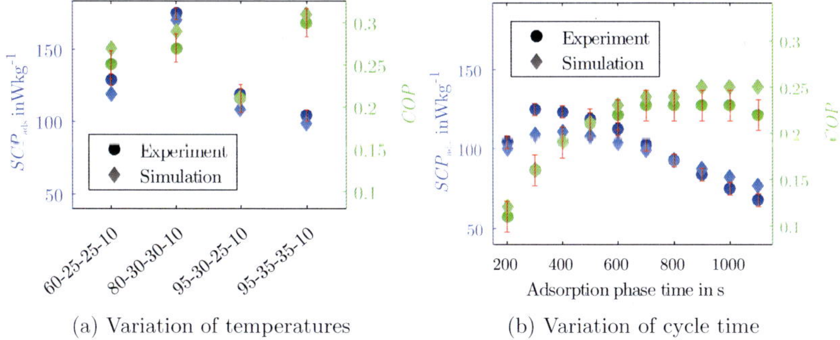

(a) Variation of temperatures (b) Variation of cycle time

Figure 9.6: SCP_{ads} (blue) and COP (green) of non-linear process model for different operating conditions. Left: four temperature combinations for desorption, adsorption, condensation and vaporisation (T_{Des}-T_{Ads}-T_{C}-T_{E} in °C). Phase times ($t_{\text{ads}} = 500\,\text{s}$, $t_{\text{des}} = 300\,\text{s}$) are kept constant for this variation. Right: increasing adsorption times from 200 seconds to 1100 seconds. The ratio between adsorption and desorption is 5/3. The inlet water temperatures are constant (95-35-35-10).

9.1.3 Controller: Phase time optimisation and state estimation

The actual controller consists of two parts: a state estimation and a phase time optimisation part (cf. Figure 9.1). The sequence of a single MPC loop is illustrated in Figure 9.7. In a first step, a moving horizon estimator uses measurement data of a fixed historic time horizon to predict the current system state. In a second step, the estimated state is used to solve an optimal control problem (OCP) for a fixed prediction time horizon. After converging or reaching the maximum number of iterations, the phase times of the plant are updated. In this section, the moving horizon estimator (MHE) and the underlying optimal control problem (OCP) are described.

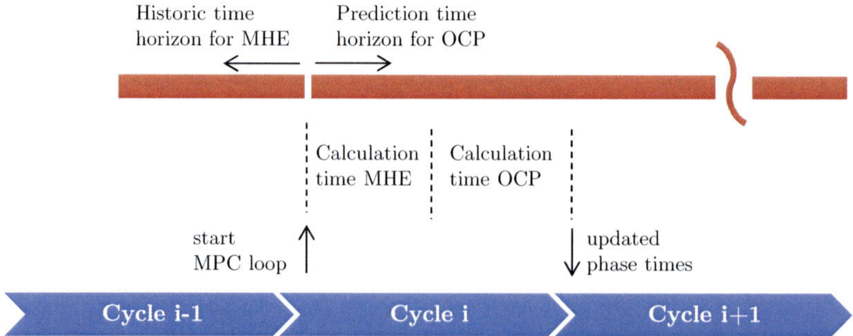

Figure 9.7: Sequence of a model-predictive-control loop: The moving horizon estimator (MHE) estimates the current system state $\hat{x}_{\mathrm{MHE}}(\tau)$ based on measurement data of a fixed historic time horizon. This state is used as initial state in an optimal control problem (OCP), determining the optimal phase times for the prediction time horizon. After solving the OCP, the updated phase times are fed back to the plant and the MPC loop starts again.

State estimator: moving horizon estimator

The investigated dynamic model is highly non-linear and includes inequality constraints, such as non-negative temperatures and loading. Therefore, a moving horizon estimator is used as a possibility to find good and robust state estimates $\hat{x}_{\mathrm{MHE}}(\tau)$ for these types of systems (Rawlings and Bakshi, 2006). A MHE solves an online optimisation problem which uses historical measurement data on a finite time horizon. In contrast to the OCP which optimises the future control, the MHE tries to find the initial states which lead to the least deviation between measured and simulated data. The resulting optimisation problem for the MHE used in this thesis reads:

$$\min_{x,z} \quad \frac{CV_{\mathrm{E}}(x(\cdot), z(\cdot), u(\cdot), p) + CV_{\mathrm{C}}(x(\cdot), z(\cdot), u(\cdot), p)}{2} \tag{9.3a}$$

$$\text{s.t.} \quad \dot{x}(t) = f^{(s)}(x(t), z(t), u(t), p)t \in \left[t^{(s)}, t^{(s+1)}\right], \tag{9.3b}$$

$$0 = g^{(s)}(x(t), z(t), u(t), p)t \in \left[t^{(s)}, t^{(s+1)}\right]. \tag{9.3c}$$

For the MHE, the considered historic time horizon influences the estimation accuracy and the optimisation time. To investigate this influence, the following test procedure is used: First, the calibration cycle is simulated with measured input data to obtain the trajectory for each state variable. Afterwards, the state variables are estimated for different historic time horizons, using the same sensor data as would be provided by the experiment (cf. Figure 9.2). The accuracy achieved by the MHE is measured with a relative state deviation (RSD), taking into account the two most difficult states to estimate, adsorbent loading w and adsorbent temperature T_{ad}:

$$RSD = \frac{w_{\mathrm{MHE}} - w_{\mathrm{sim}}}{w_{\mathrm{sim}}} + \frac{T_{\mathrm{ad,MHE}} - T_{\mathrm{ad,sim}}}{T_{\mathrm{ad,sim}}}. \tag{9.4}$$

The obtained results are shown in Figure 9.8. With increasing historic time horizon, the relative deviation of states decreases, from $RSD \approx 3\,\%$ for $t_{\mathrm{MHE}} = 200\,\mathrm{s}$ to $RSD \approx 1\,\%$ for $t_{\mathrm{MHE}} = 800\,\mathrm{s}$. At the same time, the calculation time to conduct the estimation increases from $t_{\mathrm{calc}} = 140\,\mathrm{s}$ to $t_{\mathrm{calc}} = 1362\,\mathrm{s}$. Since calculation speed of the MHE directly affects the update time of the control, an historic time horizon of $t_{\mathrm{MHE}} = 300\,\mathrm{s}$ is chosen for all further investigations. For this time horizon, the calculation time is relatively low, while achieving a state deviation of $RSD \approx 2\,\%$.

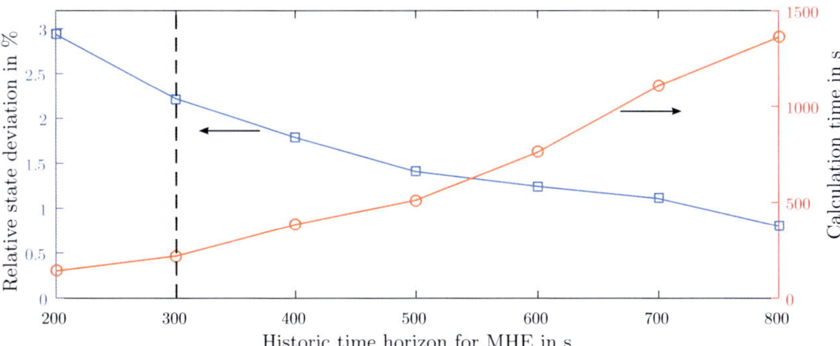

Figure 9.8: Relative state deviation (RSD) (blue, □) and calculation time (red, ∘) of moving horizon estimator with changing historic time horizon. A higher accuracy in state estimation (left axis) leads to an increase in calculation time needed for state estimation (right axis). A historic time horizon of 300 s is chosen (indicated as dashed line).

Phase time optimisation: optimal control problem

The optimal control problem is formulated similar to Equation (5.10), but without applying the cyclic steady-state constraint. Instead, an initial-value constraint is added which uses the recent state estimate $\hat{x}_{\mathrm{MHE}}(\tau)$ for the physical time τ, obtained from a moving horizon state estimator (MHE).

$$\min_{x,z,t^{(j)}} \quad -SCP_{\mathrm{ads}}(x(\cdot), z(\cdot), u(\cdot), p) \qquad \text{(Objective function)} \tag{9.5a}$$

$$\text{s.t.} \quad \dot{x}(t) = f^{(s)}(x(t), z(t), u(t), p) t \in [t^{(s)}, t^{(s+1)}] \,, \qquad \text{(Dynamic model)} \tag{9.5b}$$
$$0 = g^{(s)}(x(t), z(t), u(t), p) t \in [t^{(s)}, t^{(s+1)}], \tag{9.5c}$$
$$0 = x(t^{(0)}) - \hat{x}_{\text{MHE}}(\tau) \,, \qquad \text{(Initial value)} \tag{9.5d}$$

For the experimental investigations conducted in this thesis, only SCP_{ads} is used as objective function. For further investigations, also COP could be considered for the model predictive control.

The described optimal control problem (Equation (9.5)) is again solved using the multiple shooting algorithm `MUSCOD-II` (see Section 5.2.3).

One degree of freedom of the OCP is the time horizon used for prediction. This time horizon is not fixed to a constant time, but to a constant number of cycles. To find a suitable cycle number, the OCP is solved for a successively decreasing cycle number. For six cycles, the optimal phase times vary in the range of 2 s when moving to seven cycles. Since 2 s is the measurement interval of the plant, the prediction time horizon of the OCP is set to six cycles.

Predicted future inputs

To optimise the control for future operation, a prediction of the future input data is necessary. In the presented case of an adsorption chiller, the future inputs are the inlet temperatures $T_{\text{fl,E,in}}$, $T_{\text{fl,C,in}}$, and $T_{\text{fl,A,in}}$, the ambient temperature T_{ambient}, and the inlet volume flow rates $\dot{V}_{\text{E}}$, $\dot{V}_{\text{C}}$, and $\dot{V}_{\text{A}}$ (cf. Figure 9.2). Except for the adsorber inlet temperature $\dot{T}_{\text{fl,A,in}}$, all input conditions are reasonably constant (cf. Figure D.3). Therefore, these inlet conditions are predicted to be constant in the near future, based on the last historic measurement data. The adsorber inlet temperature $\dot{T}_{\text{fl,A,in}}$ has a step change when switching between adsorption and desorption phase. After this step, the adsorber inlet temperature approaches the set temperature. Additionally, the adsorber inlet temperature shows two fluctuations shortly after each step which is caused by the periphery of our test stand. Due to the connection between the process model in Modelica and the optimisation software MUSCOD II in C++, it has not been possible to use a set of functions to accurately approximate the adsorber inlet temperature. Therefore, the adsorber tube is replaced by a heat capacity and an additional heat resistance $(\alpha A)_{\text{set}}$ as shown in Figure 9.3. This model modification allows to approximate the adsorber inlet temperature using only the set temperature $T_{\text{Ads,set}}$ and $T_{\text{Des,set}}$ which are available from historic measurement data. For more information on predicting the future inputs see Appendix D.

9.2 Heat-flow-based controls

In contrast to fixed phase times, heuristic controls determine phase times based on a defined criterion. Based on heuristic controls proposed in literature, in this section, three novel control strategies for one-bed chillers are proposed.

Schicktanz (2011) and Büttner and Mittelbach (2013) proposed to use the chilled water outlet temperature of the evaporator $T_{\mathrm{fl,E,out}}$ for control: when the average chilled water outlet temperature $\overline{T}_{\mathrm{fl,E,out}}$ in a cycle drops below a certain setpoint $T_{\mathrm{fl,E,out,set}}$, adsorption is terminated.

Gräber (2013) proposed to use the cooling power in the evaporator $\dot{Q}_{\mathrm{E}}$ as criterion. From the starting point of maximizing the cooling power,

$$\min_{t_{\mathrm{ads}}} \quad \tfrac{1}{t_{\mathrm{ads}}} \int_0^{t_{\mathrm{ads}}} -\dot{Q}_{\mathrm{E}}(t) \, \mathrm{d}t \quad , \tag{9.6}$$

he derived that adsorption should be terminated when the current cooling power $\dot{Q}_{\mathrm{E}}$ drops below the average cooling power $\overline{\dot{Q}}_{\mathrm{E}} = \frac{1}{t_{\mathrm{ads}}} \int_0^{t_{\mathrm{ads}}} \dot{Q}_{\mathrm{E}}(t) \, \mathrm{d}t$. This criterion seems intuitive since the average cooling power starts to decrease when the current cooling power drops below the average cooling power. Gräber (2013) uses two major assumptions when deriving his control heuristic:

(1) He assumes equal cycle times for adsorption and desorption. This assumptions is justified since he considers a two-bed chiller with one evaporator and one condenser as described in Chapter 8. For this configuration, equal phase times are common practice (although also for this configuration independent phase times are beneficial as shown in Chapter 8). Based on the assumption of equal phase times, it is sufficient to define only one criterion to terminate adsorption.

(2) He assumes that cooling power during adsorption is not influenced by the time for desorption. This assumption is true when desorption characteristics are much faster than adsorption characteristics, as could be observed for silica gel (cf. Chapters 5 - 7).

Gräber (2013) shows that for a two-bed adsorption chiller with silica gel this heuristic leads to a stable control which quickly adapts to changing temperature inputs.

In the following Section 9.2.1, the main idea of Gräber (2013) is used to derive new heat-flow-based control strategies for one-bed chillers. While developed for one-bed chillers, it has to be mentioned that the derived control strategies can also be used for modularly-built multi-bed adsorption chillers based on multiple one-bed configurations, e. g. Núñez et al. (2007). In Section 9.2.2, the performance of the new heat-flow-based control strategies is tested using dynamic simulations.

9.2.1 Novel heat-flow-based control strategies for one-bed configurations

In a first step, the control strategy proposed by Gräber is adopted for a one-bed chiller by keeping the equal cycle time assumption (Control strategy 1). In a second step, a criterion to determine an independent desorption phase time is proposed leading to a (near) Pareto-optimal operation (Control strategy 2). Finally, the second control strategy is further enhanced to also allow for (near) Pareto-optimal part-load operation (Control strategy 3). The three control strategies are tested for silica gel 123 using the simulation models described in Chapter 6 - 7.

Control strategy 1 (C1) is following the criterion of Gräber: when the current evaporator heat flow rate drops below the average evaporator heat flow rate, the adsorption phase is terminated. The desorption phase time is set equal to the adsorption phase time. This leads to the following criterion:

$$\frac{1}{t_{\mathrm{des}}+t_{\mathrm{ads}}}\int_0^{t_{\mathrm{des}}+t_{\mathrm{ads}}}\dot{Q}_{\mathrm{E}}(t)=\dot{Q}_{\mathrm{E}}(t)\,, \tag{9.7a}$$

$$\frac{\mathrm{d}\dot{Q}_{\mathrm{E}}(t)}{\mathrm{d}t}<0\,, \tag{9.7b}$$

$$t_{\mathrm{ads}}=t_{\mathrm{des}}\,. \tag{9.7c}$$

Control strategy 2 (C2) is eliminating the constraint of equal adsorption and desorption phase times. Instead, a similar criterion as for vaporisation is applied for condensation: When the current condenser heat flow rate drops below the average condenser heat flow rate, the desorption phase is terminated. This criterion aims at using the desorption time as efficiently as possible:

$$\frac{1}{t_{\mathrm{des}}+t_{\mathrm{ads}}}\int_0^{t_{\mathrm{des}}+t_{\mathrm{ads}}}\dot{Q}_{\mathrm{E}}(t)=\dot{Q}_{\mathrm{E}}(t)\,, \tag{9.8a}$$

$$\frac{\mathrm{d}\dot{Q}_{\mathrm{E}}(t)}{\mathrm{d}t}<0\,, \tag{9.8b}$$

$$\frac{1}{t_{\mathrm{ads}}+t_{\mathrm{des}}}\int_0^{t_{\mathrm{ads}}+t_{\mathrm{des}}}\dot{Q}_{\mathrm{C}}(t)=\dot{Q}_{\mathrm{C}}(t)\,, \tag{9.8c}$$

$$\frac{\mathrm{d}\dot{Q}_{\mathrm{C}}(t)}{\mathrm{d}t}<0\,. \tag{9.8d}$$

The last proposed control strategy for the one-bed chiller configuration (C3) aims at providing a certain cooling load as efficiently as possible. For this purpose, the cooling power criterion is replaced: when the average cooling power drops below a certain cooling power setpoint $\overline{\dot{Q}}_{\mathrm{E,set}}$, adsorption phase is terminated. To ensure a high COP, long cycle times are desired. Thus, it is also demanded that the average cooling power has to be falling again:

$$\frac{1}{t_{\text{des}} + t_{\text{ads}}} \int_0^{t_{\text{des}}+t_{\text{ads}}} \dot{Q}_{\text{E}}(t) = \overline{\dot{Q}}_{\text{E,set}} \,, \tag{9.9a}$$

$$\frac{\text{d}}{\text{d}t} \left(\frac{1}{t_{\text{des}} + t_{\text{ads}}} \int_0^{t_{\text{des}}+t_{\text{ads}}} \dot{Q}_{\text{E}}(t) \right) < 0 \,, \tag{9.9b}$$

$$\frac{1}{t_{\text{ads}} + t_{\text{des}}} \int_0^{t_{\text{ads}}+t_{\text{des}}} \dot{Q}_{\text{C}}(t) = \dot{Q}_{\text{C}}(t) \,, \tag{9.9c}$$

$$\frac{\text{d}\dot{Q}_{\text{C}}(t)}{\text{d}t} < 0 \,. \tag{9.9d}$$

It has to be noted that, when the cooling power setpoint $\overline{\dot{Q}}_{\text{E,set}}$ is larger than the maximum achievable cooling power, Control 3 converges to Control 2. Thus, Control 3 can be used for part load and for full load. Nevertheless, Control 2 and 3 are named differently in the further investigations to enhance readability.

9.2.2 Simulation-based performance test

In this section, the three proposed heat-flow-based control strategies are tested using the dynamic simulation models described in Chapter 6 with the initial design proposed by Lanzerath (2014). For this design, the optimal cycle times are already determined (cf. Figure 6.4) and serve as benchmark for the new heat-flow-based control strategies. First, the cyclic steady-state behaviour is investigated. Afterwards, the step response of Control 2 for a change of the evaporator inlet temperature is investigated.

Cyclic steady-state performance

Figure 9.9 shows the heat flow rates and the loading for the three control strategies for one cycle in cyclic steady state. It can be observed that Control 1 is not optimal due to the fixed ratio between adsorption and desorption: the loading trajectory for Control 1 shows that desorption reaches equilibrium after 8 min, but that the desorption phase time is around 14 min. Control 2 terminates the desorption phase faster, after around 7 min. The shorter desorption time leads to a faster increase of the average cooling power and, thus, also the adsorption phase time is shortened. These shorter phase times lead to a smaller loading difference for Control 2. Control 3 also terminates the desorption phase after around 8 min. Due to the fixed cooling power setpoint ($\overline{\dot{Q}}_{\text{E,set}} = 350\,\text{W}$ in Figure 9.9), the adsorption phase time is longer than for Control 2. This leads to the largest loading difference in this case.

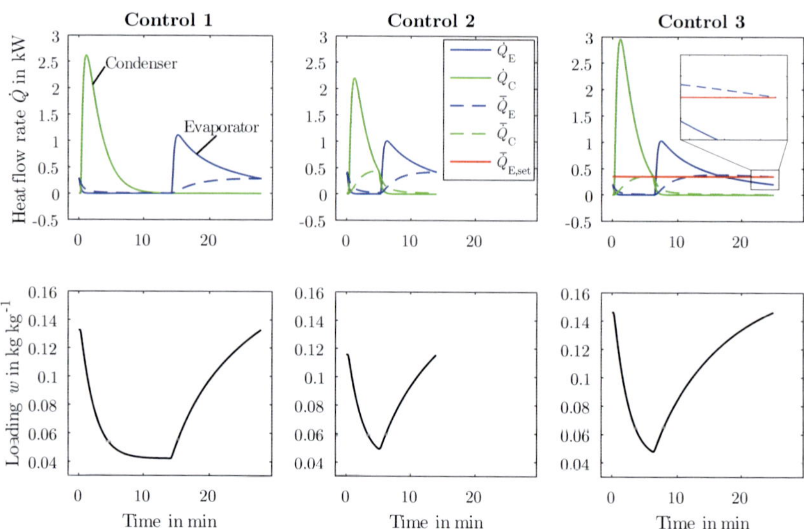

Figure 9.9: Comparison of heat-flow-based control strategies for one-bed adsorption chillers in cyclic steady state. Current and average heat flow rates of evaporator and condenser (top) and loading of the adsorber (bottom) are shown. Left: Control 1 ($\dot{Q}_{\mathrm{E}} = \bar{\dot{Q}}_{\mathrm{E}}$, $t_{\mathrm{ads}} = t_{\mathrm{des}}$); Middle: Control 2 ($\dot{Q}_{\mathrm{E}} = \bar{\dot{Q}}_{\mathrm{E}}$, $\dot{Q}_{\mathrm{C}} = \bar{\dot{Q}}_{\mathrm{C}}$); Right: Control 3 ($\dot{Q}_{\mathrm{E}} = \bar{\dot{Q}}_{\mathrm{E,set}} = 350\,\mathrm{W}$, $\dot{Q}_{\mathrm{C}} = \bar{\dot{Q}}_{\mathrm{C}}$).

Figure 9.10 shows the performance of the heat-flow-based control strategies compared to the Pareto frontier obtained in Chapter 6. It can be observed that Control 2 reaches a point close to the Pareto frontier. The achieved SCP_{ads} is around 6 % below the maximum achievable power density. Control 1 has a higher COP compared to Control 2, but a loss in SCP_{ads} of 32 %.

An interesting behaviour can be observed for Control 3. Control 3 allows to operate the chiller close to the Pareto frontier, when reducing the cooling power setpoint below the maximum cooling power achieved by Control 2. Thus, Control 3 can be used to experimentally determine the Pareto frontier without individually testing pairs of adsorption/desorption times. This significantly reduces the effort to experimentally determine Pareto frontiers for designs (cf. Figure 2.12). Additionally, Control 3 can be used during operation to Pareto-optimally meet a certain cooling load.

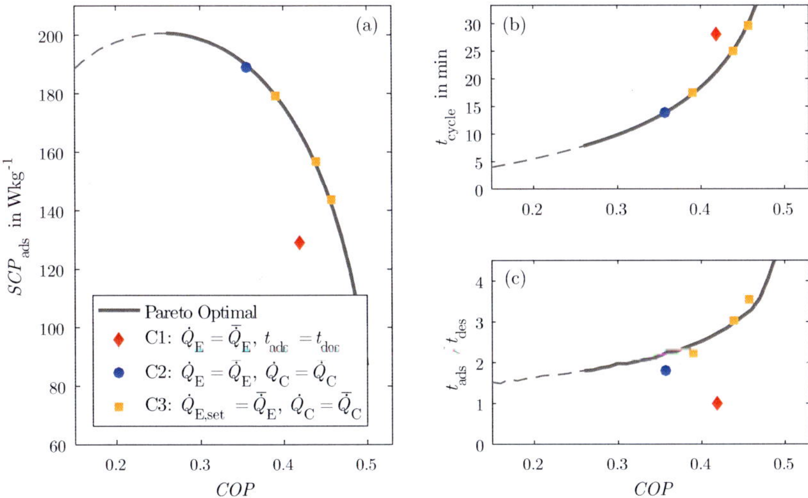

Figure 9.10: Comparison of heat-flow-based control strategies with optimal control: (a) Trade–off between specific cooling power (SCP_{ads}) and coefficient of performance (COP). (b) Optimal overall cycle times ($t_{ads} + t_{des}$). (c) Optimal adsorption / desorption phase time ratio (t_{ads}/t_{des}).

Step response

In the preceding section, the cyclic steady-state performance of the three proposed control heuristics has been investigated. In this section, the step response of Control 2 is investigated to determine how fast the control reacts: Figure 9.11 shows the heat flow rates and the loading when changing the evaporator inlet temperature from $T_{fl,E,in} = 10\,°C$ to $T_{fl,E,in} = 16\,°C$. The step occurs after 33 min which is indicated by the peak in evaporator power. The control only needs one cycle to adjust to the new input conditions. Also no undesired termination of the adsorption phase is detected. The undesired termination has also been reported by Schicktanz (2011) for an increased evaporator inlet temperature when using a temperature-based control strategy. Such an undesired early termination leads to a drop in SCP_{ads} and COP.

The simulation-based analysis of the control strategies has shown that Control 2 and 3 achieve (near) Pareto-optimal performances. The step response test has shown that Control 2 has a stable behaviour and moves to the new cyclic steady state within one cycle.

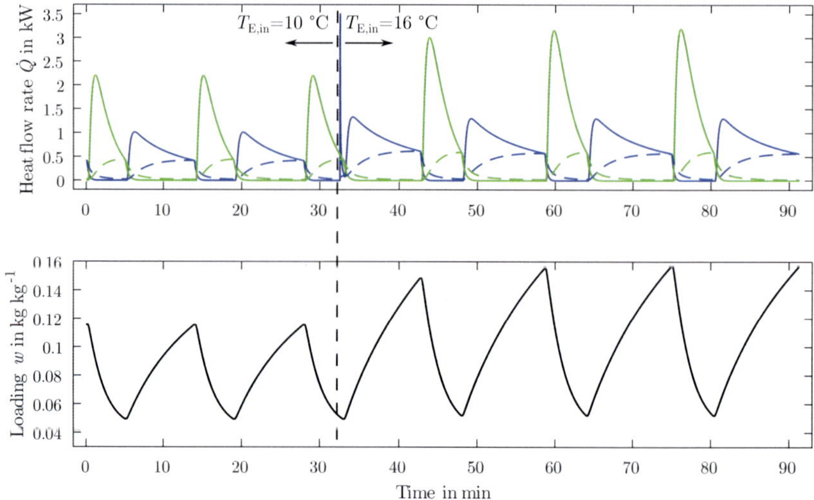

Figure 9.11: Heat flow rates and loading for simulated a step response of heat-flow-based Control strategy 2: (top) heat flow rates in evaporator and condenser, (bottom) adsorber loading. The evaporator inlet temperature changes from 10 °C to 16 °C at 33 min. The change is indicated by a dashed line.

In the following section, MPC and control strategies 1 and 2 are experimentally tested for a step response and a solar cooling test case.

9.3 Experimental validation of control strategies for one-bed chillers

In the preceding two sections, the MPC and 3 heat-flow-based control strategies have been presented. The non-linear process model of the MPC has been validated and the heat-flow-based control strategies have been tested with simulations. In this section, MPC and heat-flow-based control strategies are experimentally investigated using a modular adsorption chiller test bench (cf. Appendix E): First, in Section 9.3.1, the control strategies are investigated for several step changes in inlet temperatures of adsorber and evaporator. Second, in Section 9.3.2, the control strategies are tested for a 6 hours solar cooling test case with varying input temperatures of condenser and adsorber.

9.3.1 Experimental validation: step response

The experimental step response test aims at answering two questions: (1) How well do the control strategies perform in cyclic steady state for different inlet temperatures? (2) How fast do the control strategies react onto a change in inlet temperatures?

To answer these questions, for the MPC and Control 2 ($\dot{Q}_E = \bar{\dot{Q}}_E$, $\dot{Q}_C = \bar{\dot{Q}}_C$)), four temperature sets (T_{Des}-T_{Ads}-T_C-T_E in °C) are investigated: 70-30-30-10, 80-30-30-10, 90-30-30-10, and 90-30-30-16. For Control 1 ($\dot{Q}_E = \bar{\dot{Q}}_E$, $t_{ads} = t_{des}$), three temperature sets are measured due to availability of the test stand: 80-30-30-10, 90-30-30-10, and 90-30-30-16. Also, only four cycles at the temperature set 90-30-30-10 could be measured for Control 1. Figure 9.12 exemplarily shows the inlet temperatures of adsorber, condenser, and evaporator for each temperature set measured during the test of the MPC. Each step is induced during the adsorption phase of the cycle.

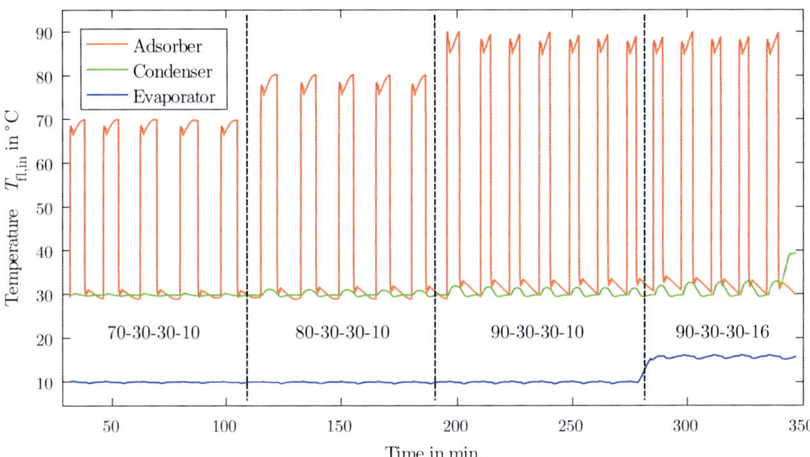

Figure 9.12: Heat exchanger inlet temperatures during a step-response measurement. The adsorber heat exchanger switches between T_{high} and T_{mid}. Condenser and evaporator have a constant temperature level during a cycle (T_{mid} and T_{low}). Each step of a temperature level is indicated by a dashed line.

Figure 9.13 exemplarily shows the heat flow rates of evaporator and condenser for a step-response measurement. The same trends as in the Chapters 5 - 8 can be observed:

The heat flow rate of the condenser is higher than the heat flow rate of the evaporator, which indicates an optimal phase time ratio $t_{\mathrm{ads}}/t_{\mathrm{des}} > 1$. This difference in heat flow rates gets more significant when moving to higher desorption temperatures $T_{\mathrm{high}} = 70\,°\mathrm{C} \rightarrow T_{\mathrm{high}} = 90\,°\mathrm{C}$. Thus, a shift in the optimal phase time ratio is to be expected. When moving to higher evaporator temperatures $T_{\mathrm{low}} = 10\,°\mathrm{C} \rightarrow T_{\mathrm{low}} = 16\,°\mathrm{C}$, both condenser and evaporator heat flow rates increase. Therefore, it is difficult to estimate the effects on control for this step.

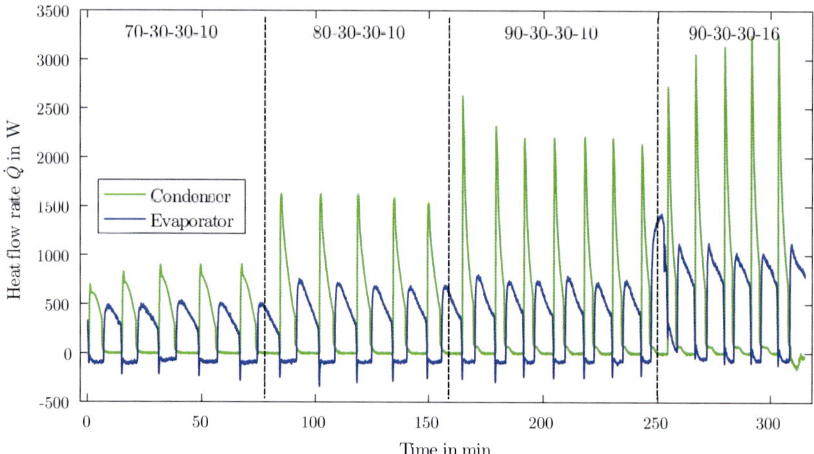

Figure 9.13: Heat flow rates of condenser and evaporator during a step-response measurement.

Figure 9.14 shows the results of the step-response measurements for the 3 control strategies: Control 1 ($\dot{Q}_{\mathrm{E}} = \overline{\dot{Q}}_{\mathrm{E}}$, $t_{\mathrm{ads}} = t_{\mathrm{des}}$), Control 2 ($\dot{Q}_{\mathrm{E}} = \overline{\dot{Q}}_{\mathrm{E}}$, $\dot{Q}_{\mathrm{C}} = \overline{\dot{Q}}_{\mathrm{C}}$), and MPC. For comparison, the power density SCP_{ads}, the efficiency COP, the cycle time t_{cycle}, and the phase time ratio $t_{\mathrm{ads}}/t_{\mathrm{des}}$ are evaluated for each cycle. For Control 1, the four cycles at 90-30-30-10 are split to better illustrate the step response. In the following sections, first, the general performance obtained by the control strategies is discussed, followed by an analysis of the response times for each control strategy.

Power density

For all 3 control strategies, the power density increases for every step (cf. Figure 9.14). This increase in SCP_{ads} is caused by a growing difference in adsorption potentials:

an increased desorption temperature enhances the maximum adsorption potential, an increased evaporator temperature lowers the minimum adsorption potential (cf. Figure 5.11). The larger difference in adsorption potentials leads to higher driving forces and, thus, to higher SCP_{ads}-values.

Comparing the control strategies regarding power densities, it can be observed that the control strategies with independent adsorption and desorption phase times (MPC and Control 2: $\dot{Q}_{\text{E}} = \overline{\dot{Q}}_{\text{E}}$, $\dot{Q}_{\text{C}} = \overline{\dot{Q}}_{\text{C}}$) outperform the control strategy with equal phase times (Control 1: $\dot{Q}_{\text{E}} = \overline{\dot{Q}}_{\text{E}}$, $t_{\text{ads}} = t_{\text{des}}$). This result confirms the optimisation results obtained in Chapters 5 - 7. Additionally to the equal phase times, Control 1 also suffers from longer cycle times compared to the other control strategies. Due to these two reasons, MPC and Control 2 achieve almost twice the power density compared to Control 1. MPC and Control 2 achieve almost equal power densities for all operating conditions (Figure 9.14 (a)), although MPC and Control 2 vary partly in the control parameters: Both control strategies lead to a similar ratio of adsorption to desorption times, but Control 2 tends to longer cycle times. These longer cycle times are beneficial regarding efficiency as discussed in the following.

Efficiency

Regarding efficiency measured by the COP, Control 2 shows the best performance for all operating conditions. The efficiency achieved by Control 1 is slightly below the values of Control 2, but outperforms the MPC for most of the operating points. These results show the positive effect of long cycle times on COP. It can be concluded that the results from the simulation study in Section 9.2.2 are confirmed: Control 2 ($\dot{Q}_{\text{E}} = \overline{\dot{Q}}_{\text{E}}$, $\dot{Q}_{\text{C}} = \overline{\dot{Q}}_{\text{C}}$) outperforms Control 1 ($\dot{Q}_{\text{E}} = \overline{\dot{Q}}_{\text{E}}$, $t_{\text{ads}} = t_{\text{des}}$) and leads to a (near) Pareto-optimal operation with high power densities and efficiencies. With the given non-linear process model, the MPC, given maximum power density as objective function, tends to shorter cycles than Control 2. This leads to similar (maximum) power densities, but lower efficiencies. One way to force the MPC to longer cycle times would be an additional constraint demanding a minimum COP. This would lead to higher efficiencies, but is not further investigated in this thesis.

Response time

Control 2 ($\dot{Q}_{\text{E}} = \overline{\dot{Q}}_{\text{E}}$, $\dot{Q}_{\text{C}} = \overline{\dot{Q}}_{\text{C}}$) shows the fastest responses given the tested changes in inlet temperatures. After only one cycle, Control 2 approaches the new steady-state control parameters.

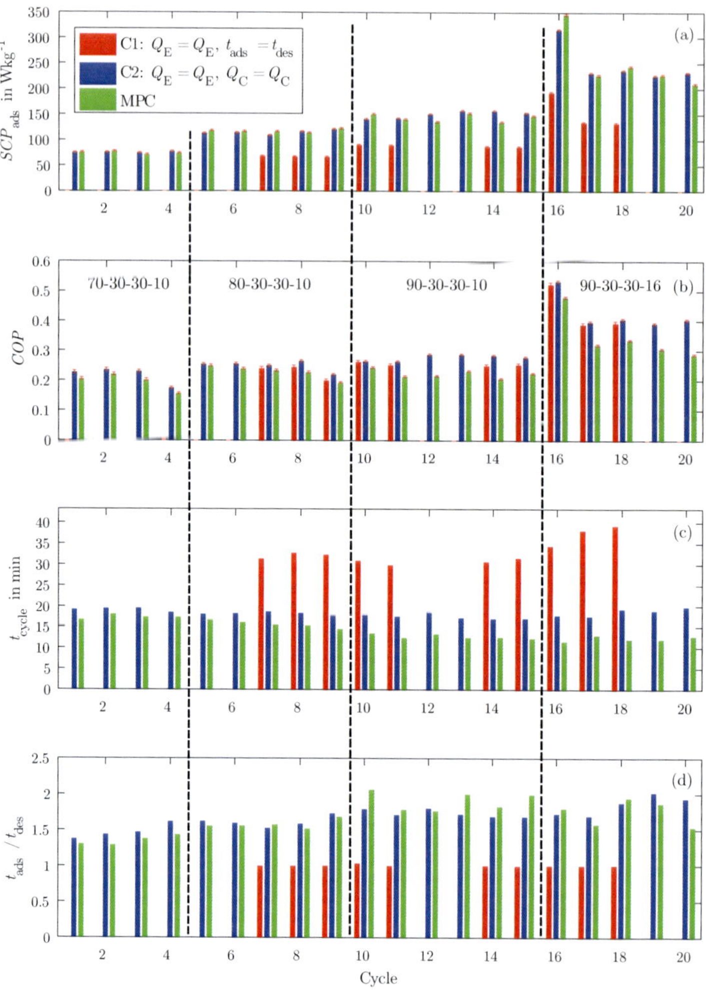

Figure 9.14: Performance and control during the step-response measurements for three control strategies: MPC, Control 1 ($\dot{Q}_E = \overline{\dot{Q}}_E$, $t_{ads} = t_{des}$), and Control 2 ($\dot{Q}_E = \overline{\dot{Q}}_E$, $\dot{Q}_C = \overline{\dot{Q}}_C$). (a) Specific cooling power SCP_{ads}, (b) coefficient of performance COP, (c) cycle time t_{cycle}, and (d) phase time ratio t_{ads}/t_{des}.

Control 1 ($\dot{Q}_{\mathrm{E}} = \overline{\dot{Q}}_{\mathrm{E}}$, $t_{\mathrm{ads}} = t_{\mathrm{des}}$) does reach steady state only slowly. This can be best observed for the cycle time in Figure 9.14 (c), which moves like a wave. This slow response behaviour compared to Control 2 is caused by the equal phase time criterion. The equal phase times lead to an amplification of the control criterion in the evaporator: When the evaporator tends to longer adsorption phase times, this also leads to a longer desorption phase time. This longer desorption phase time changes the starting point for the next adsorption. Additionally, the longer desorption phase time reduces the average cooling power due to a larger cycle time. This, again, leads to a longer adsorption phase time and so on.

The MPC shows a slow but stable response speed due to the necessary time for optimisation and state estimation. Given a change in inlet conditions, the MPC steers the system to the next steady-state control parameters within 2-3 cycles. It is, therefore, well suited to react on slowly changing operating conditions, e. g. changes in weather conditions.

In summary, MPC and Control 2 ($\dot{Q}_{\mathrm{E}} = \overline{\dot{Q}}_{\mathrm{E}}$, $\dot{Q}_{\mathrm{C}} = \overline{\dot{Q}}_{\mathrm{C}}$) outperform Control 1 ($\dot{Q}_{\mathrm{E}} = \overline{\dot{Q}}_{\mathrm{E}}$, $t_{\mathrm{ads}} = t_{\mathrm{des}}$). Additionally, Control 2 shows a fast response for changing inlet conditions, whereas the MPC steers the system to the next optimal operating point. In the following section, all three control strategies are tested for continuously varying inlet conditions representing a typical solar cooling case.

9.3.2 Experimental validation: Solar cooling

The following experimental setup aims at emulating a typical solar cooling case. Assuming a rising sun, the desorption temperatures (T_{high}) increases from 60 °C to 90 °C and the ambient temperature (T_{mid}) rises from 25 °C to 35 °C. The peak is reached after 3 h. After this point, both desorption temperature and ambient temperature are falling again. The evaporator temperature is assumed to be constant at 10 °C. In total, the test case lasts 6 h. The inlet temperatures are shown in Figure 9.15.

For the MPC, the objective function is again: maximum power density. The obtained results for the 3 tested control strategies are shown in Figure 9.16. The performance is shown in Figure 9.16 (a) and (b), and the control parameters are shown in Figure 9.16 (c) and (d).

Control 2 and the MPC show a similar power density which has already been observed for the step response. Both control strategies outperform C1 during the entire cycle. Regarding *COP*, Control 1 and Control 2 show the best performance, since *COP* is not considered in the MPC in this thesis. The cycle time follows the

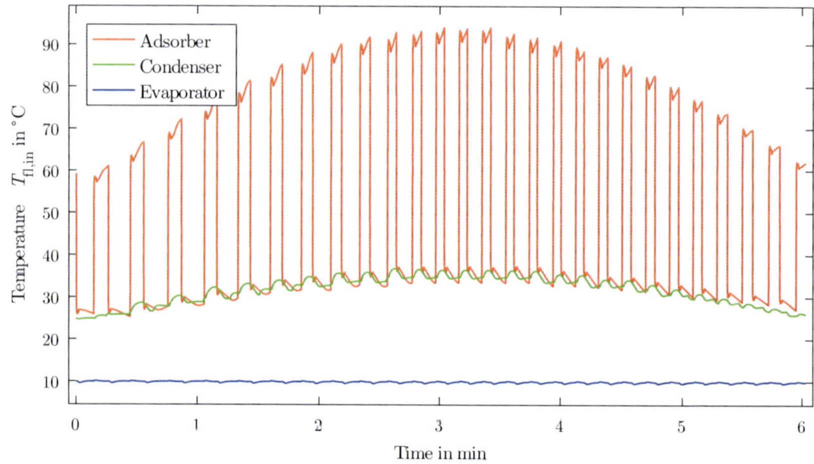

Figure 9.15: Heat exchanger inlet temperatures during a solar-cooling measurement.

inverse trend of the inlet temperatures: At 3 h, the shortest cycle times are observed for all control strategies. Afterwards, for Control 1 and Control 2, the cycle times rise again. The MPC tends to shorter cycle times and only rises slowly after the peak in desorption and ambient temperature. The phase time ratio of Control 2 and of the MPC is between 1.5 and 2 with mostly a higher phase time ratio for the MPC.

All in all, the results for the solar day show the same trend as the step response: the MPC and Control 2 achieve the same power densities, but with shorter cycle times and lower efficiencies for the MPC.

Figure 9.17 quantifies the average performance. In terms of power density, average SCP_{ads}-values of 124.2 W kg^{-1} (Control 2) and 126.6 W kg^{-1} (MPC) are achieved. Control 1 only reaches an SCP_{ads} of 90.6 W kg^{-1} which is 30 % less than for the other control strategies. In terms of efficiency, Control 1 reaches an average COP of 0.282, Control 2 of 0.2748, and the MPC of 0.238. For the MPC, this means a reduction of around 16 % compared to Control 1, keeping in mind that COP is not included as an objective for the MPC in these investigations.

The results show that independent phase times are crucial for a high power density given the investigated one-bed adsorption chiller.

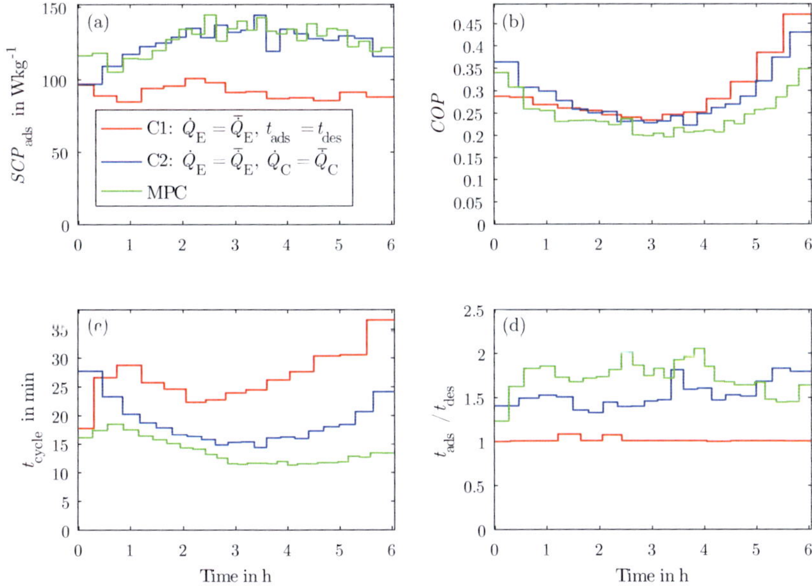

Figure 9.16: Performance and control during the solar-cooling measurements for three control strategies: Control 1 ($\dot{Q}_E = \bar{\dot{Q}}_E$, $t_{ads} = t_{des}$), Control 2 ($\dot{Q}_E = \bar{\dot{Q}}_E$, $\dot{Q}_C = \bar{\dot{Q}}_C$), and MPC. (a) Power density SCP_{ads}, (b) Coefficient of performance COP, (c) cycle time t_{cycle}, and (d) Phase time ratio t_{ads}/t_{des}.

9.4 Final remarks on optimal control of adsorption chillers

In this chapter, a model predictive control has been successfully applied for a maximum power density operation. Additionally, a heat-flow-based control strategy with independent phase times is developed and successfully tested. This heat-flow-based control strategy allows for (near) Pareto-optimal operation of one-bed adsorption chillers.

The heat-flow-based control strategy C2 and the MPC show similar performance regarding the investigated objective: maximum power density. When moving beyond one-bed chillers to more complex systems, each control strategy has its benefits which are shortly discussed in the following.

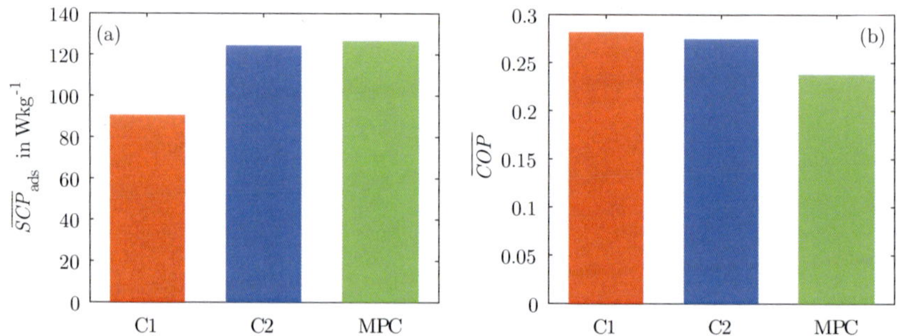

Figure 9.17: Average performance for the solar-cooling measurements for the three control strategies: Control 1 ($\dot{Q}_{\text{E}} = \overline{\dot{Q}}_{\text{E}}$, $t_{\text{ads}} = t_{\text{des}}$), Control 2 ($\dot{Q}_{\text{E}} = \overline{\dot{Q}}_{\text{E}}$, $\dot{Q}_{\text{C}} = \overline{\dot{Q}}_{\text{C}}$), and MPC. (a) Average power density $\overline{SCP}_{\text{ads}}$ and (b) Average coefficient of performance $\overline{COP}$.

Benefits of model predictive control with increasing system complexity

For the simple one-bed adsorption chillers, the MPC seems to be unnecessarily complex, since Control strategy 2 achieved similar power densities. This immediately changes when moving to more complex system configurations including more control parameters, e. g. fan speed of cooling towers, volume flow rate in the heat exchanger circuits, complex multi-bed adsorption configurations, larger energy systems consisting of adsorption chillers and compression chillers,... . For all of these systems, the control problem becomes too complex to achieve an optimal control with a simple heuristic. The experimental proof of concept of an MPC for adsorption chillers is a first step towards employing MPC for more complex adsorption-based energy systems.

Heat-flow-based (near) Pareto-optimal control strategy for commercial adsorption chillers

The heat-flow-based control strategies Control 2 and Control 3 lead to a Pareto-optimal operation of one-bed adsorption chillers. Since the control variables (heat flow rates) can be easily measured, this control strategy may be easily applied to commercial adsorption chillers.

Chapter 10

Summary, conclusions and future perspective

The aim of this thesis is to move from simulation-based analysis to simultaneous optimisation of design and control of adsorption energy systems. In the following, in Section 10.1, the thesis is summarised and the main conclusions are drawn. Afterwards, in Section 10.2, possible future research directions are discussed.

10.1 Summary and conclusions

Design of adsorption energy systems is a difficult task due to the following challenges: (1) intrinsic dynamics, (2) multi-objectiveness, (3) large variety in design parameters, (4) strong influence of control, (5) large impact of input parameters, and (6) interaction between design, control, and input parameters. In this thesis, a rigorous optimisation approach is followed to handle these challenges. A multi-objective optimal control problem has been formulated and solved. The approach is used to exemplary find the optimal design and control of a finned-tube adsorption chiller.

The major contributions and findings of this thesis are highlighted separately in the following.

Adsorption Energy Systems Library

An object-oriented dynamic-model library is developed in the modelling language Modelica. The library provides basic models of adsorption energy systems. With these basic models, new adsorption chillers, heat pumps, thermal storage systems, or desiccant systems can be easily modelled. In addition, a large number of equilibrium data is included which allows for a fast exchange of the investigated working pair. In this thesis, the dynamic-model library is used to develop the process models used for optimisation and optimal control.

Simultaneous optimisation of control and design

A multi-objective DAE-constrained optimal control problem has been formulated and solved. The optimal control problem is used to simultaneously optimise design and control of adsorption chillers. Based on a calibrated adsorption chiller model, design parameters, such as grain size, fin number, component sizing, and cycle design have been assessed and optimised while guaranteeing optimal control. Thus, the intrinsic performance of design choices is compared and not the effects of poor control.

Identifying the bottleneck – The loss in adsorption potential

Adsorption energy systems can be modelled as a series of resistances and capacities. Each resistance reduces the potential for adsorption or desorption and lowers the performance of the system. Identifying the main resistances is a first step to an optimal design. In adsorption energy systems, multiple heat and mass transfer resistances are present with different potentials, such as temperature, pressure, or loading. In this thesis, the different potentials are translated to one measure: the loss of adsorption potential. Thus, a direct comparison of heat and mass transfer resistances is possible. The method is used to identify the dominant resistances for the investigated design.

How to measure power density? – The effect on design

Adsorption chillers are commonly evaluated by two performance indicators: power density and efficiency. To determine power density, often the mass of adsorbent is used as reference measure, while neglecting the mass or volume of other components. Using the adsorbent mass as a reference for the power density, an optimisation always tends to systems with a low adsorbent mass, although total weight or volume restrictions could be more relevant. This effect is demonstrated for the optimal adsorber-bed design and component sizing, leading to a significant change of the optimal design.

Optimal cycle design? – A question of cooling load

The optimal operating strategy for two-bed chillers with different heat recovery schemes is investigated. It is observed that the overall Pareto-frontier consists of two operating strategies. Thus, depending on the cooling load, the operating strategy should be changed from a passive heat recovery scheme with independent adsorption / desorption phase times to an active heat recovery scheme with equal adsorption / desorption phase times.

Model predictive control

The dynamic-model library and the employed optimisation techniques have been used to set up a model predictive control (MPC) for adsorption chillers. The MPC allows for arbitrary objective functions and the consideration of path constraints. The MPC is successfully employed to optimise the phase times of a one-bed adsorption chiller.

Heat-flow-based control strategy for Pareto-optimal operation

In addition to the MPC, new heat-flow-based control strategies are derived for one-bed or modular multi-bed adsorption chillers. The heat-flow-based control strategies lead to a (near) Pareto-optimal operation for any given cooling load. Since the necessary input data (volume flow rates and temperatures) are easy to measure, the proposed control strategies have a high potential to improve operation of commercial adsorption chillers.

10.2 Future research

Possible future research topics can be derived based on the presented methods and findings in this thesis. In the following, these possible research topics are clustered and main questions are formulated.

Validation of optimisation results

The optimisation results in this thesis have not been validated. It would be interesting to built a modified adsorption chiller with the optimal design. This chiller should be compared to the reference design proposed by Lanzerath (2014) to quantify the gain in performance.

Creating a full design map

In this thesis, the simultaneous optimisation approach has been used to optimise a specific adsorption chiller design proposed by Lanzerath (2014). The investigation could be extended to more design choices, e. g. more working pairs, coatings, different evaporator and condenser types, and more cycle designs. By this, a full multi-objective design map could be drawn consisting of multiple Pareto-frontiers. Since input conditions are controlled, the obtained Pareto frontiers could be directly compared. Of course, this design map could also be determined for different temperature triples. By this, the best design for a certain application could be easily determined.

From small-scale experiments to optimal adsorption energy systems

In this thesis, calibrated component models are used as a starting point. These component models have been calibrated by a full-scale experiment. To reduce experimental effort, it is desirable to derive heat and mass transfer coefficients directly from small-scale experiments like the large-temperature-jump experiment. To do this, the obtained measurement information, such as characteristic times, have to be translated into heat and mass transfer coefficients for full-scale experiments. Although, first steps have been conducted (e. g. by Graf et al. (2016)), validation of the predicted performance of the full-scale chiller has not been shown yet.

Increase system boundaries – consider pumps, cooling towers, and heat sources

In this thesis, the design and control of the adsorption chiller itself has been optimised. In a next step, the system boundaries should be increased to the periphery. The periphery, such as pumps, cooling towers, or heat sources determine the inlet conditions of the adsorption chiller. It has been shown that the inlet conditions strongly influence the system performance. Since pumps and cooling towers are consuming electricity which can be incorporated as a cost, a trade-off arises between good inlet conditions and electricity consumption. This trade-off can be formulated in a new enhanced optimisation problem.

Integration of adsorption energy systems into larger energy systems

In real world applications, adsorption energy systems are often part of a larger energy system. In these larger energy systems, adsorption energy systems interact with other components, such as thermal storage units, compression chillers and heat pumps, ... It is a difficult design task to find the optimal sizing and operation of such systems, given the system dynamics and the potential degrees of freedom. The methods presented in this thesis may do their part in moving one step towards optimisation of large energy systems including adsorption technology.

Appendices

Appendix A

Working pairs

In this appendix, the used equilibrium data for silica gel 123 / water and AQSOA Z02 / water are summarised. For both working pairs, the characteristic curve and the specific heat capacity is documented.

A.1 Silica gel 123 / water

The characteristic curve for silica gel 123 / water is reported in Schawe (1999). The reported characteristic curve $W = f(A)$ is valid in a range of $A > 17\,\mathrm{kJ\,kg^{-1}}$ to $A > 1940\,\mathrm{kJ\,kg^{-1}}$:

$$W = d + \frac{a}{\pi}\left[\arctan\left(\frac{A-b}{c}\right) + \frac{\pi}{2}\right] , \tag{A.1}$$

with

$$\begin{aligned} a &= 5.072\,313 \times 10^{-1} , \\ b &= 1.305\,531 \times 10^{2} , \\ c &= -8.492\,403 \times 10^{1} , \\ d &= 4.128\,962 \times 10^{3} . \end{aligned}$$

The specific heat capacity is reported by Schawe (1999) to be:

$$c_{\mathrm{sor}} = 1000\,\mathrm{J\,kg^{-1}\,K^{-1}} . \tag{A.2}$$

A.2 AQSOA Z02 / water

The characteristic curve for AQSOA Z02 / water is obtained from measurement data by Goldsworthy (2014). Based on this measurement data, the parameters of an arctan function are fitted, minimising the squared error. Figure A.1 shows both measurement data and fit.

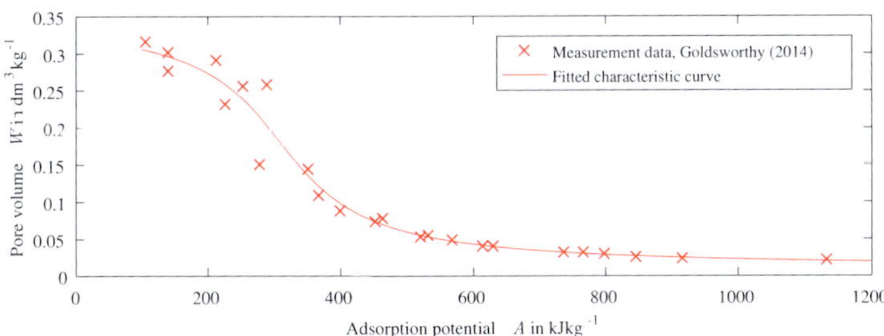

Figure A.1: Fit of characteristic curve for AQSOA Z02 / water based on measurement data by Goldsworthy (2014).

The fitted characteristic curve $W = f(A)$ is valid in a range of $A > 106\,\mathrm{kJ\,kg^{-1}}$ to $A > 1140\,\mathrm{kJ\,kg^{-1}}$:

$$W = d + \frac{a}{\pi}\left[\arctan\left(\frac{A-b}{c}\right) + \frac{\pi}{2}\right], \tag{A.3}$$

with

$$a = -3.461\,028 \times 10^{-1},$$
$$b = 3.096\,522 \times 10^{2},$$
$$c = 9.726\,662 \times 10^{1},$$
$$d = 3.535\,178 \times 10^{-1}.$$

The specific heat capacity is assumed to be similar to the one of silica gel 123 / water:

$$c_{\mathrm{sor}} = 1000\,\mathrm{J\,kg^{-1}\,K^{-1}}. \tag{A.4}$$

Appendix B

Heat exchanger discretisation

Lanzerath (2014) used discretised models for the heat exchangers (adsorber, evaporator, and condenser) to determine the heat transfer coefficients $(\alpha A)_{\mathrm{ad}}$, $(\alpha A)_{\mathrm{E}}$, and $(\alpha A)_{\mathrm{C}}$. This discretisation is used to model the plug flow in the heat exchangers as accurate as possible. The drawback of the discretisation is the high number of states. For optimisation purposes, a low discretisation is desired to keep the optimisation problem as small as possible. To compensate for the error caused by the reduced discretisation, a model modification is used which is described in this appendix.

The discretised model of an heat exchanger used by Lanzerath (2014) is illustrated in Figure B.1.

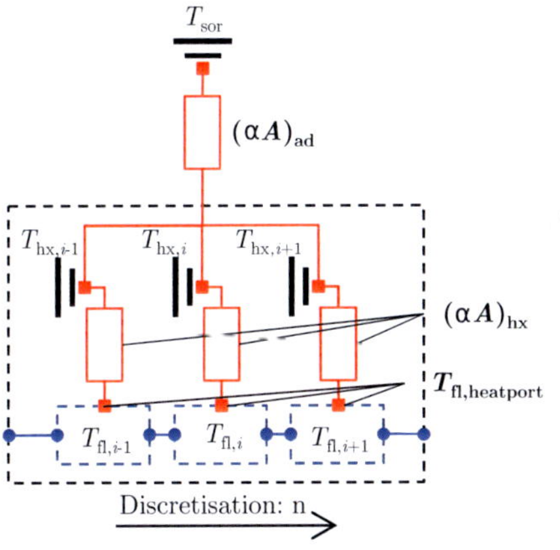

Figure B.1: Structure of discretised heat exchanger used by Lanzerath (2014), consisting of liquid cells, heat capacities, and heat transfer models.

In each liquid cell, the energy and mass balance is solved as described in Section 3.2.3. It is assumed that the liquid in each cell is well-mixed and has a homogeneous temperature $T_{\mathrm{fl},i}$, which is also the outlet temperature $T_{\mathrm{fl},i} = T_{\mathrm{fl,out},i}$.

As described in Section 3.3.1, the heat transfer between two components is determined by the temperature difference between these components ΔT and a heat transfer coefficient αA. For the liquid cell, different options exist to set the temperature used for heat flow calculation $T_{\mathrm{fl,heatport},i}$ (cf. Figure B.1). Lanzerath (2014) assumed $T_{\mathrm{fl,heatport},i}$ to be the bulk temperature $T_{\mathrm{fl},i}$. By doing this, the absolute heat flow rate is always underestimated, because the real temperature along the flow direction is higher on average. This underestimation becomes less important for high discretisations. To compensate for this underestimation, a linear temperature distribution along the flow direction is assumed:

$$T_{\mathrm{fl,heatport},i} = \frac{T_{\mathrm{fl,in},i} + T_{\mathrm{fl,out},i}}{2} = \overline{T}_{\mathrm{fl},i} \,. \tag{B.1}$$

This approach has the downside that the overall absolute heat flow rate may be overestimated, if the temperature distribution is not close to a linear distribution, e. g. at switching times or when the ratio between heat capacity flow and heat transfer coefficient is low.

In the following, the two approaches are tested for the adsorption chiller model used in Chapter 6 for two discretisions: (1) $n_{\mathrm{ads}} = 40$, $n_{\mathrm{E}} = n_{\mathrm{C}} = 10$ (used by Lanzerath (2014)), (2) $n_{\mathrm{ads}} = 1$, $n_{\mathrm{E}} = n_{\mathrm{C}} = 1$. As test case, the optimal one-bed chiller obtained in Chapter 6 is used: $d_{\mathrm{p}} = 0.375\,\mathrm{mm}$, $n_{\mathrm{fins}} = 14.7$, $l_{\mathrm{hx,E}} = 3.27\,\mathrm{m}$, and $l_{\mathrm{hx,C}} = 2.2\,\mathrm{m}$.

Figure B.2 shows the inlet and outlet temperatures for the 4 discretisation cases for adsorber, evaporator, and condenser heat exchanger. Especially in the evaporator and in the condenser, the linear temperature distribution with a discretisation of $1-1-1$ is very close to the reference case $T_{\mathrm{fl,out}}$, $40-10-10$ used by Lanzerath (2014). Since the evaporator outlet temperature is proportional to the SCP_{all}, a good agreement between $T_{\mathrm{fl,out}}$, $40-10-10$ and $\overline{T}_{\mathrm{fl}}$, $1-1-1$ is expected. If only discretisation is reduced, but the heatport temperature is not adapted ($T_{\mathrm{fl,out}}$, $1-1-1$), a large error in all heat exchanger outlet temperature can be observed.

For the linear temperature distribution with a discretisation of $1-1-1$, a peak at the outlet temperature of the adsorber can be observed (cf. Figure B.2). This non-physical peak is caused by the assumption of a linear temperature distribution in the tube, which is not valid right after switching the inlet temperature. Due to the fact that this behaviour only occurs at switching events and that the overall agreement between $T_{\mathrm{fl,out}}$, $40-10-10$ and $\overline{T}_{\mathrm{fl}}$, $1-1-1$ is very high, the non-physical behaviour can be accepted.

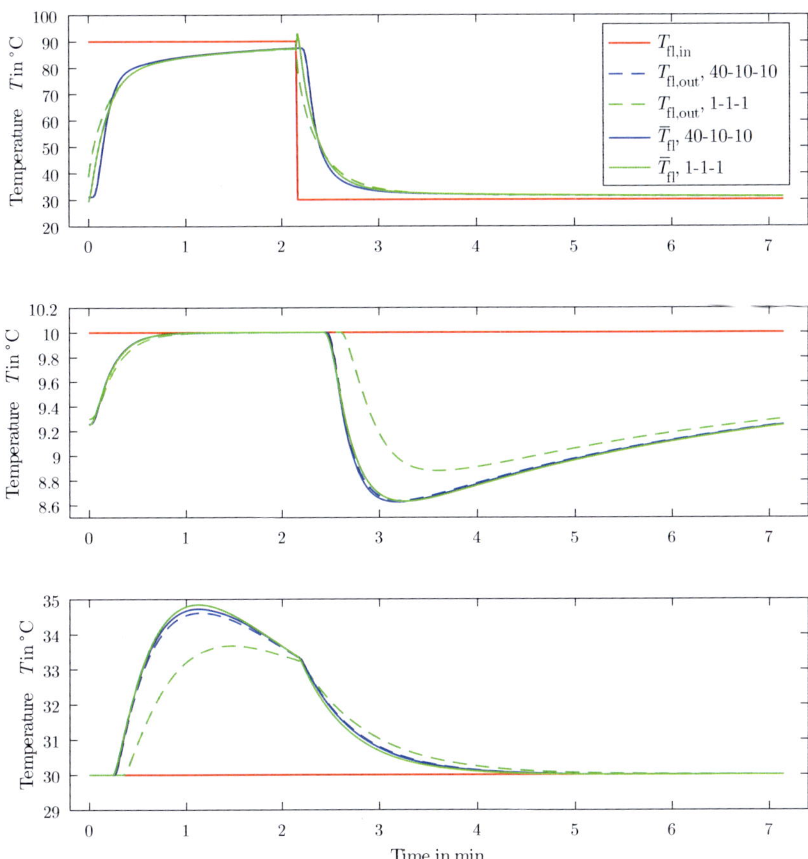

Figure B.2: Comparison of outlet temperatures for four discretisation schemes. (top). adsorber heat exchanger, (middle): evaporator heat exchanger, and (bottom): condenser heat exchanger.

Figure B.3 shows the comparison between the discretisation options regarding COP and SCP_{all}. Again, a reduction of discretisation without adapting the heatport temperature leads to a significant decrease of both COP and SCP_{all} (-8 % and -15 %). By assuming a linear temperature distribution and removing the discretisation, the difference in COP and SCP_{all} becomes very small (1 % and 0.9 %). This error is accepted since differential states are decreased from 128 to 12.

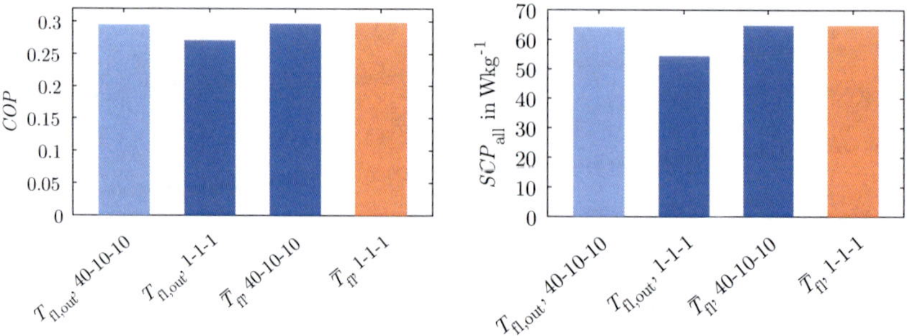

Figure B.3. Comparison of performance for four dicretisation schemes. (left): coefficient of performance COP, (right): specific cooling power SCP_{all}.

Appendix C

Two-bed cycle schemes

In this appendix, the 5 cycle designs investigated in Chapter 8 are described in detail. For each cycle design, the heat exchanger fluid connections are shown for every phase.

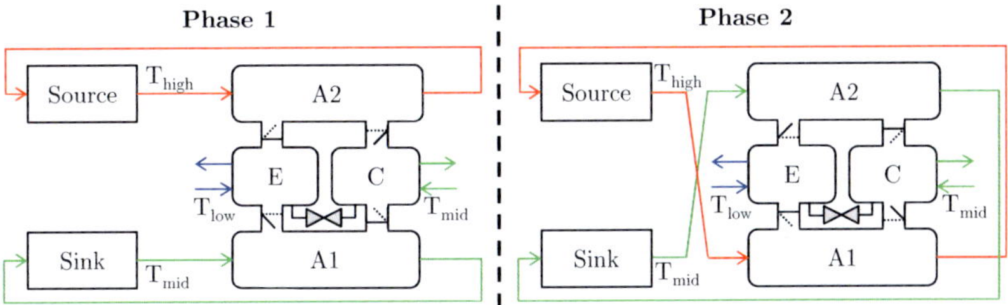

Figure C.1: Phases of simple two-bed chiller cycle with equal phase times (NoHR+EPT). The chiller consists of two adsorbers (A1 and A2), an evaporator (E), and a condenser (C). The adsorbers are connected to the evaporator and the condenser by flap valves. Condenser and evaporator are connected by a return valve. In phase 1, A1 is in adsorption mode and connected to the heat sink, A2 is in desorption mode and connected to the heat source. In phase 2, the modes for A1 and A2 are swapped.

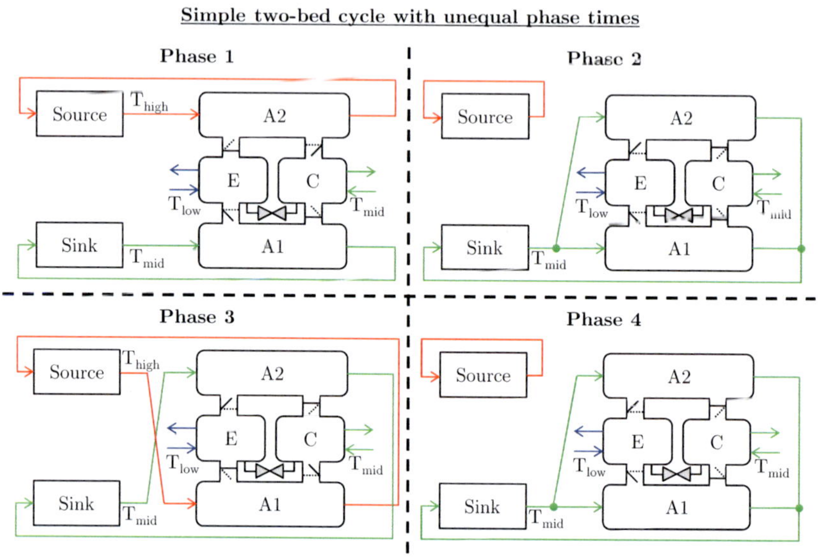

Simple two-bed cycle with unequal phase times

Figure C.2: Phases of simple two-bed chiller cycle with independent phase times (NoHR+IPT). The chiller consists of two adsorbers (A1 and A2), an evaporator (E), and a condenser (C). The adsorbers are connected to the evaporator and the condenser by flap valves. Condenser and evaporator are connected by a return valve. In phase 1, A1 is in adsorption mode and connected to the heat sink, A2 is in desorption mode and connected to the heat source. In phase 2, both adsorbers are in adsorption mode and connected to the heat sink. In phase 3, A1 is in desorption and A2 is in adsorption mode. In phase 4, again both adsorbers are in adsorption mode.

Two-bed cycle with active heat recovery and equal phase times

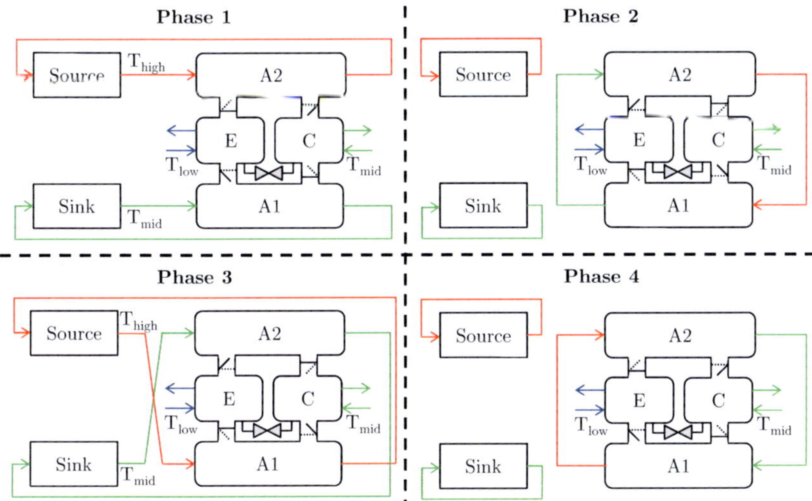

Figure C.3: Phases of two-bed chiller cycle with active heat recovery (AHR). The chiller consists of two adsorbers (A1 and A2), an evaporator (E), and a condenser (C). The adsorbers are connected to the evaporator and the condenser by flap valves. Condenser and evaporator are connected by a return valve. In phase 1, A1 is in adsorption mode and connected to the heat sink and A2 is in desorption mode and connected to the heat source. In phase 2, the adsorbers are connected and decoupled from heat sink and heat source. In phase 3, A1 is in desorption and A2 is in adsorption mode. In phase 4, the adsorbers are again connected.

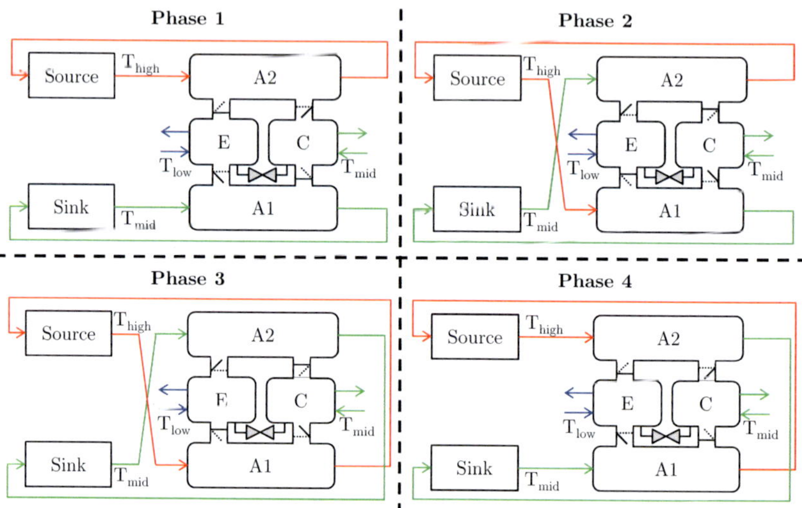

Figure C.4: Phases of two-bed chiller cycle with passive heat recovery and equal phase times (PHR+EPT). The chiller consists of two adsorbers (A1 and A2), an evaporator (E), and a condenser (C). The adsorbers are connected to the evaporator and the condenser by flap valves. Condenser and evaporator are connected by a return valve. In phase 1, A1 is in adsorption mode and connected to the heat sink, A2 is in desorption mode and connected to the heat source. In phase 2, heat source and sink at the entrance of the adsorbers are swapped, but not at the outlet. In phase 3, A1 is in desorption and A2 is in adsorption mode. In phase 4, again heat source and sink at the entrance of the adsorbers are swapped, but not at the outlet.

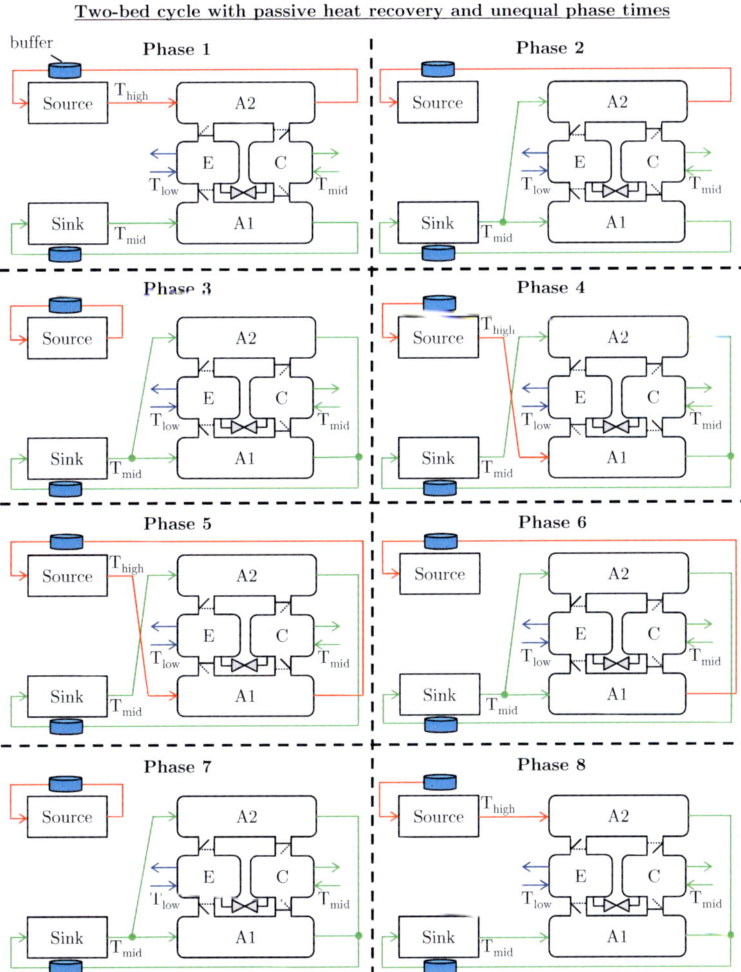

Figure C.5: Phases of two-bed chiller cycle with passive heat recovery and independent phase times (PHR+IPT). The chiller consists of two adsorbers (A1 and A2), an evaporator (E), and a condenser (C). The adsorbers are connected to the evaporator and the condenser by flap valves. Condenser and evaporator are connected by a return valve. Phases 1, 3, 5, and 7 are the phases of the simple cycle with independent phase times (cf. Figure C.2). Phases 2, 4, 6, and 8 are the additional phases needed for heat recovery of the internal energy of the heat exchanger fluid (cf. Figure C.4).

Appendix D

Model modifications and calibration of non-linear process model for MPC

In this appendix, the final process model used for the model predictive control is derived stepwise: First, a modified version of the process model used in Chapter 6 is presented. Second, this process model is altered to incorporate future external inputs by simple means.

In principle, the process model is similar to the model described in Chapter 6 with three modifications: (1) heat losses are added, (2) inter-particle mass transfer resistance (Darcy) is neglected because grain sizes of $d_p = 0.9\,\mathrm{mm}$ are used (cf. Section 5.3.3), (3) a pressure loss due to friction in the butterfly valves is added ($\beta_{\mathrm{friction}}$). Figure D.1 shows the model structure of the modified process model. Since the model is changed and also the evaporator and condenser are changed compared to Part II, a new calibration of the process model is necessary.

Model calibration is conducted stepwise according to the procedures proposed by Lanzerath (2014) and Schreiber (2017): First, the heat loss coefficients $(\alpha A)_{\mathrm{loss,C}}$, $(\alpha A)_{\mathrm{loss,E}}$, $(\alpha A)_{\mathrm{loss,A}}$, and $(\alpha A)_{\mathrm{casing,A}}$ are determined from steady-state measurements. Afterwards, a cyclic measurement is used to determine all missing parameters. The evaporator and condenser heat transfer coefficients, $(\alpha A)_{\mathrm{E}}$ and $(\alpha A)_{\mathrm{C}}$, are calibrated with sub-models. Afterwards, the friction parameter is determined by pressure measurements in the adsorber, evaporator and condenser. Finally, the adsorber heat transfer coefficient $(\alpha A)_{\mathrm{ad}}$ and the intra-particle mass transfer coefficient β_{LDF} are determined for adsorption and desorption.

As criterion for accuracy, the coefficient of variation regarding the measured $\dot{Q}_{\mathrm{meas,i}}$

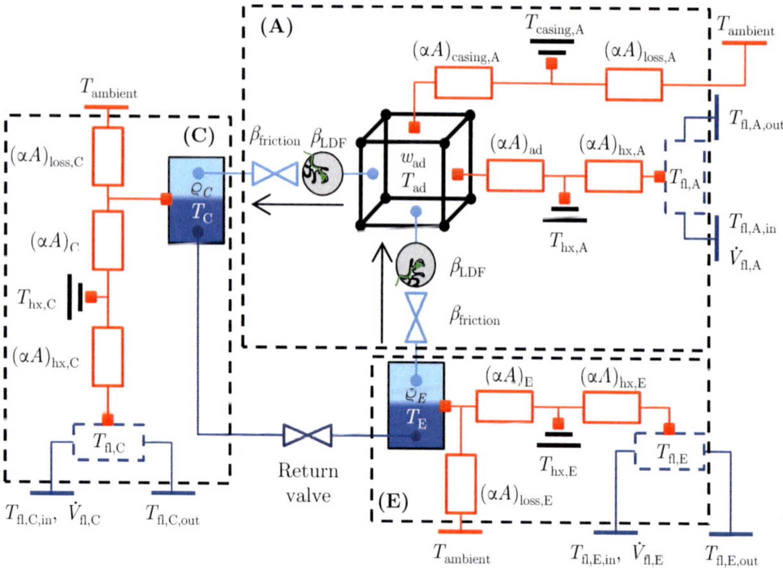

Figure D.1: Structure of the non-linear process model used for MPC. The process model consists of three components: evaporator (E), condenser (C), and adsorber (A). Heat transfer coefficients and differential states are named.

and simulated heat flow rates $\dot{Q}_{\text{sim,i}}$ is used:

$$CV_{\text{i}} = \frac{1}{n}\sum_{i=1}^{n}\frac{\sqrt{\frac{1}{t_{\text{cycle}}}\cdot\int(\dot{Q}_{\text{meas,i}}-\dot{Q}_{\text{sim,i}})^2\text{d}t}}{\bar{\dot{Q}}_{\text{meas,i}}}. \tag{D.1}$$

In this thesis, the heat flow rates of evaporator and condenser are used.

For calibration, the following temperatures and cycle times are used: $T_{\text{high}} = 95\,°\text{C}$, $T_{\text{mid}} = 30\,°\text{C}$, $T_{\text{low}} = 10\,°\text{C}$, $t_{\text{ads}} = 450\,\text{s}$, and $t_{\text{des}} = 300\,\text{s}$. The obtained model parameters for the calibration cycle are listed in Table D.1. The achieved accuracy is shown in Figure D.2. Both, evaporator and condenser heat flow rates show good agreement between measured and simulated data.

In principle, this calibrated modified process model would be suitable for model predictive control as described in Section 9.1, but due to implementation reasons, an additional modification is necessary. For the model predictive control, a prediction concerning the future inputs is necessary (cf. Figure 9.1 and Section 9.1.3). Figure D.3

Table D.1: Fitted parameters of the modified non-linear process model.

Parameter		
$\beta_{\mathrm{LDF,ads}}$	5.93e-3	s^{-1}
$\beta_{\mathrm{LDF,des}}$	2.30e-2	s^{-1}
αA_{ad}	$(184 + 3 \cdot T_{\mathrm{ad}})$	$\mathrm{W\,K}^{-1}$
αA_{E}	2156	$\mathrm{W\,K}^{-1}$
αA_{C}	2179	$\mathrm{W\,K}^{-1}$
$\alpha A_{\mathrm{casing,A}}$	6	$\mathrm{W\,K}^{-1}$
$\alpha A_{\mathrm{loss,A}}$	1.2	$\mathrm{W\,K}^{-1}$
$\alpha A_{\mathrm{loss,C}}$	1	$\mathrm{W\,K}^{-1}$
$\alpha A_{\mathrm{loss,E}}$	1.5	$\mathrm{W\,K}^{-1}$

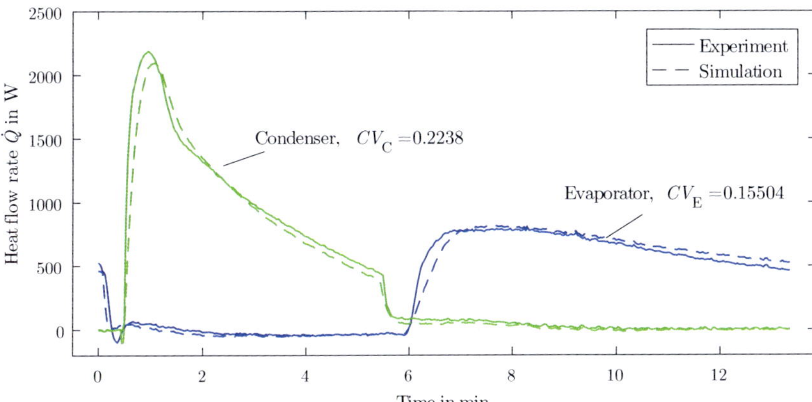

Figure D.2: Simulation accuracy of the non-linear process model. The simulated (−−) and measured (−) heat flow rates of evaporator and condenser are shown.

shows the inputs of the calibration cycle. Due to the connection between the process model in Modelica and the optimisation software MUSCOD II in C++, it has been not possible to use a set of functions to accurately approximate the adsorber inlet temperature. Therefore, the adsorber tube is replaced by a heat capacity and an additional heat resistance as shown in Figure 9.3.

This model modification allows to approximate the adsorber inlet temperature using only the set temperature $T_{\mathrm{Ads,set}}$ and $T_{\mathrm{Des,set}}$ which are available from historic meas-

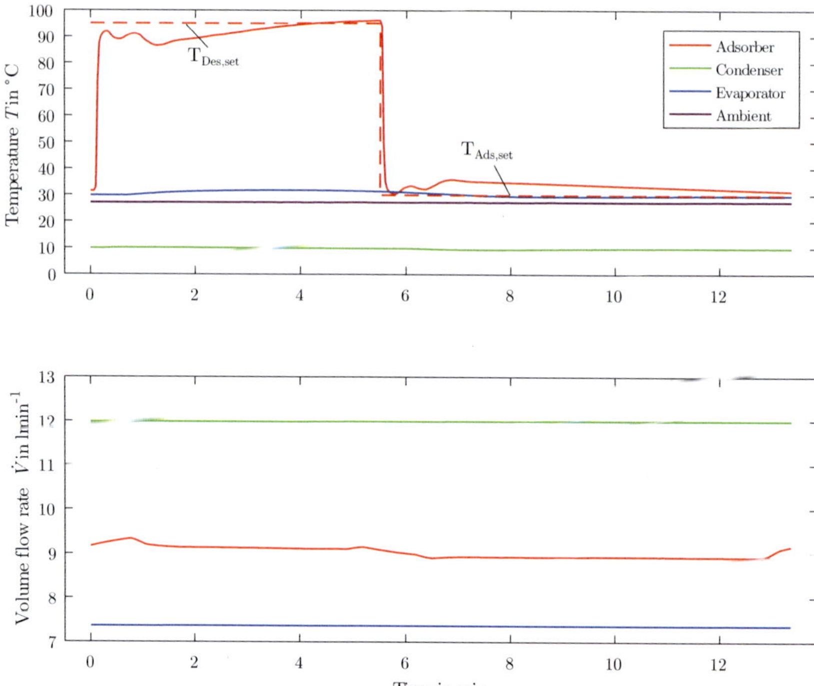

Figure D.3: Inlet conditions of calibration cycle: temperatures and volume flow rates. For the adsorber inlet temperature $T_{\mathrm{fl,A,in}}$, also the set temperatures for adsorption $T_{\mathrm{Ads,set}}$ and desorption $T_{\mathrm{Des,set}}$ are shown.

urement data. This is done by varying the heat resistance $(\alpha A)_{\mathrm{set}}$ which leads to a sequence of exponential functions with varying slope. Figure D.4 shows the approximated adsorber inlet conditions using this approach. Although, the two fluctuations at the beginning of each step do not show a perfect fit, the overall characteristic of the adsorber inlet temperature is captured well by the modified model. The final model as shown in Figure 9.3 is validated in Section 9.1.2.

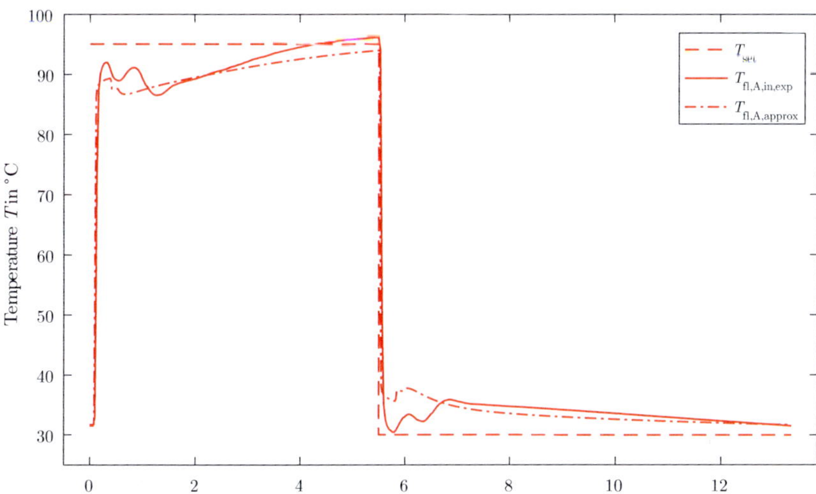

Figure D.4: Comparison of adsorber inlet temperatures for calibration: set temperature T_{set}, experimental inlet conditions $T_{fl,A,in,exp}$, and approximated inlet conditions using heat capacity modification $T_{fl,A,approx}$.

Appendix E

Test bench and components

In this appendix, the most important information on the test bench and the components used in Chapter 9 are provided. The test bench allows to experimentally evaluate adsorption chillers. In this thesis, a one-bed modularly-built adsorption chiller is investigated. The general set-up of the test-bench and the chiller is described by Lanzerath (2014). Figure E.1 shows a picture of the adsorption chiller.

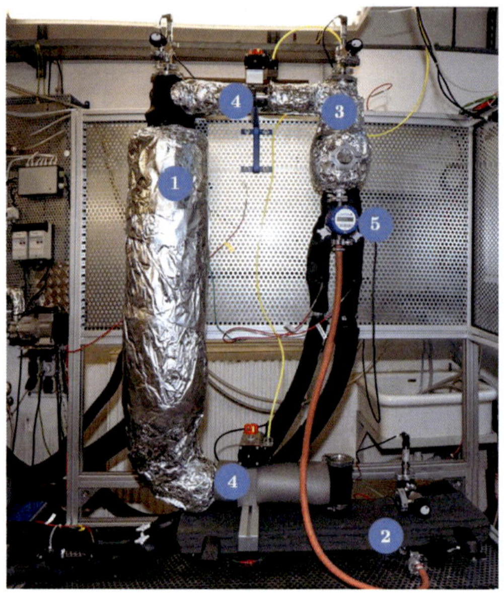

Figure E.1: Picture of modular adsorption-chiller test bench: (1) adsorber, (2) evaporator, (3) condenser, (4) butterfly valve, and (5) return valve.

E.1 Components

In this section, the components used for experimental evaluation of the control strategies are presented.

E.1.1 Adsorber

The adsorber is the same as used for the rigorous optimisation study (Chapters 5 - 8): The adsorber consists of 13 tubes. Each tube is 460 mm, where 400 mm are covered with fins (Figure E.2a). The inner diameter of the tube is 13.5 mm, the outer diameter is 36 mm. All data is summarised in Table E.1.

Table E.1: Geometrical data of adsorber heat exchanger

Parameter		Value	
Total length heat exchanger	l_{hx}	6.8	m
Inner finned surface area	A_{hx}	0.45	m^2
Outer finned surface area	A_{ad}	2.15	m^2
Mass water	m_{fl}	0.866	kg
Mass aluminum	$m_{hx,1}$	3.946	kg
Mass stainless steel	$m_{hx,2}$	0.338	kg
Mass silica gel 123	m_{sor}	2.236	kg
Heat capacity ratio (metal/sorbent)	C_{hx}/C_{sor}	1.64	

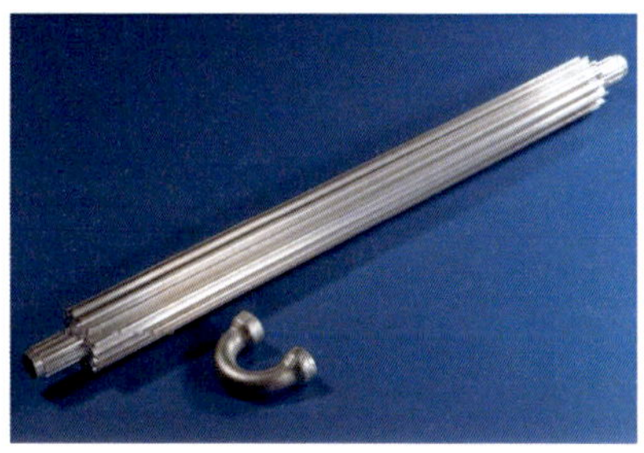

(a) Heat exchanger tube adsorber

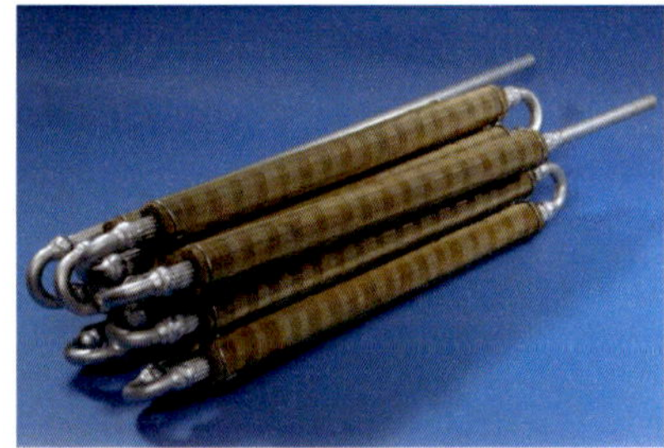

(b) Heat exchanger package adsorber

Figure E.2: Adsorber: finned annulus-tube heat exchanger. (a) Close-up of finned heat exchanger tube. (b) heat exchanger package

E.1.2 Evaporator

For the experimental study in Chapter 9, a micro-fin evaporator is used. The tubes are of the type GEWA-K-4009.17090-00. The evaporator consists of 15 tubes with a tube length of 650 mm each. The inner diameter of the tube is 15.2 mm and the equivalent outer tube is 17.8 mm. Figure E.3 shows the tubes and the casing of the evaporator. Table E.2 summarises the evaporator data.

Table E.2: Geometrical data of evaporator heat exchanger

Parameter		Value	
Total length heat exchanger	l_{hx}	9.75	m
Inner finned surface area	A_{hx}	0.47	m^2
Outer finned surface area	A_{E}	0.55	m^2
Mass water	m_{fl}	1.769	kg
Mass copper	m_{hx}	5.85	kg

(a) Tubes

(b) Casing

Figure E.3: Micro-fin evaporator. (a) tubes, (b) casing

E.1.3 Condenser

The condenser is built as a double-helix heat exchanger. The copper tube has an inner diameter of 8 mm and an outer diameter of 10 mm. Figure E.4 shows the double helix and the casing. The main condenser data are summarised in Table E.3.

Table E.3: Geometrical data of condenser heat exchanger.

Parameter		Value	
Total length heat exchanger	l_{hx}	4.625	m
Inner surface area	A_{hx}	0.116	m^2
Outer surface area	A_{C}	0.145	m^2
Mass water	m_{fl}	0.232	kg
Mass copper	m_{hx}	1.167	kg

(a) Double-helix tube

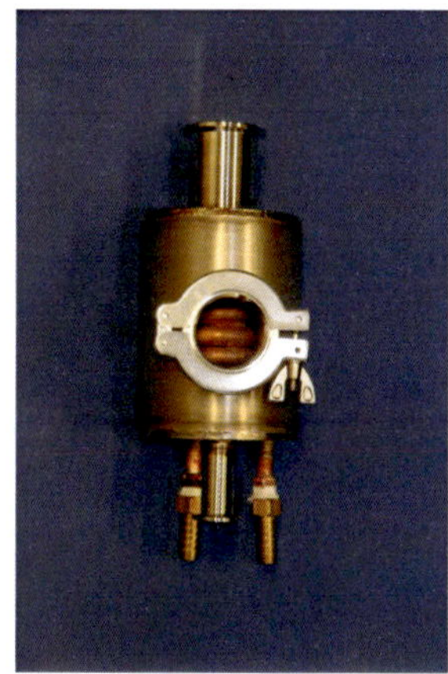

(b) Casing

Figure E.4: Double-helix condenser. (a) tube, (b) casing

E.1.4 Valves and connections

The adsorber is connected to the evaporator and the condenser by a buterfly valve each. The butterfly valves are opened and closed by a pneumatic impulse. This ensures that the valves can be opened/closed also for high pressure differences. All components are connected via KF40 or KF16 fittings.

E.1.5 Thermal bathes

Four thermal bathes are used to supply the heat exchanger fluid inlet temperatures.

Table E.4: Summary of used thermal baths.

Component	Manufacturer	Type	Heating power	Cooling power
Adsorber (heat source at T_{high})	Single	STW 200 /1-12-32-HO	12 kW	41.8 kW
Adsorber (heat sink at T_{mid})	Single	STW 200 /1-12-32-HO.2	12 kW	41.8 kW
Condenser	Thermo Scientific	G50/PC200	2 kW	1 kW
Evaporator	Thermo Scientific	G50/PC200	2 kW	1 kW

E.1.6 Measurement equipment and uncertainty

The performance of the adsorption chiller is monitored by temperature, pressure and volume flow rate sensors. To allow for interpretation of the results, knowledge about the measurement uncertainty is crucial. The measurement uncertainty is obtained differently for each type of measurement equipment. All uncertainties are reported according to the "Guide to the expression of uncertainty in measurement" Joint Committee for Guides in Metrology (2008) as standard uncertainty u with a coverage factor $k = 1$. Table E.5 summarises the uncertainties for all sensors used (compare Figure 9.2).

For more information on the general measurement equipment see Lanzerath (2014).

E.2 Operation

The test bench is operated in 6 phases, corresponding in general to the 4 phases of the simple cycle explained in Section 2.1. Compared to the simple cycle, isosteric heating and cooling are each separated into two phases, pre-heating + heating and pre-cooling + cooling respectively. This phase separation allows recovering the internal energy of the fluid in the adsorber heat exchanger which reduces the heating and cooling demand at the thermal bathes. The position of each valve in every phase is summarised in Table E.6.

Table E.5: Overview on used sensors and maximum uncertainty in the range of operation.

Measurement	Sensor	Variable	Uncertainty	
Temperature	PT100	$T_{\mathrm{E,in}}$	0.023	K
Temperature	PT100	$T_{\mathrm{E,out}}$	0.023	K
Temperature	PT100	$T_{\mathrm{C,in}}$	0.023	K
Temperature	PT100	$T_{\mathrm{C,out}}$	0.023	K
Temperature	PT100	$T_{\mathrm{A,in}}$	0.023	K
Temperature	PT100	$T_{\mathrm{A,out}}$	0.023	K
Temperature	PT100	T_{ambient}	0.023	K
Temperature	PT100	$T_{\mathrm{reservoir}}$	0.023	K
Temperature	PT100	$T_{\mathrm{E,water}}$	0.023	K
Temperature	PT100	$T_{\mathrm{E,vapour}}$	0.023	K
Temperature	PT100	$T_{\mathrm{A,vapour,in}}$	0.023	K
Temperature	PT100	$T_{\mathrm{A,vapour,out}}$	0.023	K
Temperature	TC	$T_{\mathrm{E,casing}}$	0.1	K
Temperature	TC	$T_{\mathrm{C,casing}}$	0.1	K
Temperature	TC	$T_{\mathrm{A,casing}}$	0.1	K
Pressure	Piezo-resistive	p_{E}	4.62	mbar
Pressure	Piezo-resistive	p_{C}	4.62	mbar
Pressure	Piezo-resistive	p_{A}	4.62	mbar
Volume flow rate	Vortex	$\dot{V}_{\mathrm{E}}$	0.053	$\mathrm{l\,min^{-1}}$
Volume flow rate	Vortex	$\dot{V}_{\mathrm{C}}$	0.053	$\mathrm{l\,min^{-1}}$
Volume flow rate	Magnetic	$\dot{V}_{\mathrm{A}}$	0.019	$\mathrm{l\,min^{-1}}$
Temperature difference	PT100	$T_{\mathrm{E,in}} - T_{\mathrm{E,out}}$	0.0078	K
Temperature difference	PT100	$T_{\mathrm{C,in}} - T_{\mathrm{C,out}}$	0.0013	K

Table E.6: Valve positions for the six phases. The valve names correspond to Figure 9.2.

#	Phase	Valves of heat exchanger fluid circuit V1	V2	V3	V4	Butterfly valves E/A	A/C	Return valve
0	Heating	close	close	open	open	close	close	close
1	Desorption	close	close	open	open	close	open	open
2	Pre-cooling	open	close	close	open	close	close	close
3	Cooling	open	open	close	close	close	close	close
4	Adsorption	open	open	close	close	open	close	close
5	Pre-heating	close	open	open	close	close	close	close

Appendix F

Publications and student theses

List of publications

Journal papers

Bau, U., Hoseinpoori, P., Graf, S., Schreiber, H., Lanzerath, F., Kirches, C., and Bardow, A. (2017). Dynamic optimisation of adsorber-bed designs ensuring optimal control. *Applied Thermal Engineering*, 125:1565-1576.

Erdogan, M., Graf, S., Bau, U., Lanzerath, F., and Bardow, A. (2017). Simple two-step assessment of novel adsorbents for drying: The trade-off between adsorber size and drying time. *Applied Thermal Engineering*, 125:1075-1082.

Bau, U., Braatz, A.-L., Lanzerath, F., Herty, M., and Bardow, A. (2015). Control of adsorption chillers by a gradient descent method for optimal cycle time allocation. *International Journal of Refrigeration*, 56:52-64.

Lanzerath, F., Bau, U., Seiler, J., and Bardow, A. (2015). Optimal design of adsorption chillers based on a validated dynamic object-oriented model. *Science and Technology for the Built Environment*, 21(3):248-257.

Conference contributions

Bau, U., Gibelhaus, A., Lanzerath, F., and Bardow, A. (2017). Optimal operation of two-bed adsorption chillers by combining heat recovery and reallocation of adsorption/desorption times. In *Proceedings of International Sorption Heat Pump Conference 2017*.

Bau, U., Lanzerath, F., and Bardow, A. (2017). A simple heat-flow-based control strategy for one-bed adsorption chillers leading to Pareto-optimal performance. In *Proceedings of International Sorption Heat Pump Conference 2017.*

Gibelhaus, A., Fidorra, N., Lanzerath, F., Bau, U., Schnabel, L., Köhler, J., and Bardow, A. (2017). Hybrid refrigeration with CO2 vapor compression cycle and adsorption chiller: An efficient combination of natural working fluids. In *Proceedings of International Sorption Heat Pump Conference 2017.*

Graf, S., Bau, U., Erdogan, M., and Bardow, A. (2017). The optimal adsorbent for adsorption chiller cycles. In *Proceedings of International Sorption Heat Pump Conference 2017.*

Bau, U., Schreiber, H., Lanzerath, F., and Bardow, A. (2017). Hybrid Air-Conditioning for Electric Vehicles by Combining a Heating and a Desiccant System. In *Proceedings of 26th Aachener Kolloquium*, (accepted).

Bau, U., Hoseinpoori, P., Graf, S., Schreiber, H., Lanzerath, F., and Bardow, A. (2017). Rigorous assessment of adsorber-bed designs using dynamic optimization. In Eames, I. W. and Tierney, M.J., editors, *Proceedings of Heat Powered Cycles 2016.*

Schreiber, H., Bau, U., Lanzerath, F., and Bardow, A. (2016). Heat loss in adsorption thermal energy storage: modeling and experimental validation. In Eames, I. W. and Tierney, M.J., editors, *Proceedings of Heat Powered Cycles 2016.*

Bau, U., Schreiber, H., Lanzerath, F., and Bardow, A. (2016). Hybride Klimatisierung für Elektrofahrzeuge mit Wärmepumpe und offenem Sorptionssystem (in German). In Deutscher Kälte- und Klimatechnischer Verein (DKV) e.V., editor, *DKV-Tagung 2016.*

Bau, U., Schreiber, H., Lanzerath, F., and Bardow, A. (2016). Hybride Klimatisierung für Elektrofahrzeuge: Kombination von Wärmepumpe und offenem Sorptionssystem (in German). In Deußen, N. and Steinberg, P., editors, *Wärmemanagement des Kraftfahrzeugs*, Haus-der-Technik Fachbuch.

Bau, U., Neitzke, D., Lanzerath, F., and Bardow, A. (2015). Multi-objective optimization of dynamic systems combining genetic algorithms and Modelica: Application to adsorption air-conditioning systems. In Fritzson, P. and Elmquist, H., editors, *Proceedings of the 11th International Modelica Conference*, Linköping electronic conference proceedings, pages 777-784, Linköping. Modelica Association.

Bau, U., Braatz, A.-L., Lanzerath, F., Herty, M., and Bardow, A. (2015). Optimal Dynamic Operation of Adsorption-based Energy Systems Driven by Fluctuating Renewable. In Gernaey, K. V., Huusom, J. K., and Gani, R., editors, *12th International Symposium on Process Systems Engineering and 25th European Symposium on Computer Aided Process Engineering*, volume 37 of *Computer-aided chemical engineering*, pages 2339-2344. Elsevier, Amsterdam.

Bau, U., Schreiber, H., Lanzerath, F., and Bardow, A. (2015). Adsorption-based air-conditioning for battery-driven electric busses. In *Proceedings of the 24th IIR International Congress of Refrigeration*, Paris. International Institute of Refrigeration.

Bau, U., Lanzerath, F., Gräber, M., Schreiber, H., Thielen, N., and Bardow, A. (2014). Adsorption energy systems library - Modeling adsorption based chillers, heat pumps, thermal storages and desiccant systems. In Tummescheit, H. and Arzén, K.-E., editors, *Proceedings of the 10th International Modelica Conference*, Linköping electronic conference proceedings, pages 875-883, Linköping. Modelica Association.

Lanzerath, F., Seiler, J., Bau, U., and Bardow, A. (2014). A modular experimental and simulation approach for the systematic development of adsorption heat pumps. In Radermacher, R., editor, *International Sorption Heat Pump Conference (ISHPC 2014)*, pages 117-126. Curran, Red Hook, NY.

Student theses supervised during this work

Balzer, C. (2014). Konzeptionierung und dynamische Modellierung eines Trocknungsprozesses mit verbesserter Energieeffizienz durch Integration eines Adsorbers und einer Wärmepumpe (in German). Master's thesis, RWTH Aachen University.

Braatz, A.-L. (2014). Regelung gewöhnlicher Differentialgleichungen am Beispiel einer energieeffzienten Steuerung einer wärmegetriebenen Adsorptionskälteanlage (in German). Master's thesis, RWTH Aachen University.

Baumgärtner, N. (2016). Implementation and experimental validation of a model predictive control for adsorption heat pumps. Master's thesis, RWTH Aachen University.

Engelpracht, M. (2015). Regelung, Inbetriebnahme und Bewertung eines thermischen Adsorptionsspeichers zur Klimatisierung von Elektrobussen (in German). Bachelor's thesis, RWTH Aachen University.

Ganz, K. (2015). Simulation eines Hybridsystems aus Wärmepumpe und Adsorptionsspeicher zur effizienten Klimatisierung von Elektrofahrzeugen (in German). Bachelor's thesis, RWTH Aachen University.

Hahn, J. (2014). Dynamische Modellierung eines offenen Adsorptionssystems mittels FEM (in German). Bachelor's thesis, RWTH Aachen University.

Hoseinpoori, P. (2016). Assessment and Fair Comparison of Adsorber Bed Design Employing Dynamic Optimization. Master's thesis, Politecnico Di Torino.

Neitzke, D. (2015). Modellgestützte Auslegung und Optimierung einer sorptionsgestützten Klimatisierung für Elektrobusse (in German). Master's thesis, RWTH Aachen University.

Olmedo, L. E. (2014). Dimensionality reduction of a dynamical adsorption chiller model towards linear optimization. Master's thesis, Ecole Polytechnique Fédérale de Lausanne.

Schumacher, L. (2013). Untersuchung von Wärmeumwandlungsprozessen zur Nutzung von Niedertemperaturwärme für die Kälte- und Wärmebereitstellung (in German). Mini thesis, RWTH Aachen University.

Thielen, N. (2013). Dynamische Modellierung eines offenen Adsorptions-(Luft-)Trocknungsprozesses (in German). Diploma thesis, RWTH Aachen University.

Bibliography

Akahira, A., Alam, K. C. A., Hamamoto, Y., Akisawa, A., and Kashiwagi, T. (2005a). Experimental investigation of mass recovery adsorption refrigeration cycle. *International Journal of Refrigeration*, 28(4):565–572.

Akahira, A., Alam, K. C. A., Hamamoto, Y., Akisawa, A., and Kashiwagi, T. (2005b). Mass recovery four-bed adsorption refrigeration cycle with energy cascading. *Applied Thermal Engineering*, 25(11-12):1764–1778.

Alam, K. C. A., Akahira, A., Hamamoto, Y., Akisawa, A., and Kashiwagi, T. (2004). A four-bed mass recovery adsorption refrigeration cycle driven by low temperature waste/renewable heat source. *Renewable Energy*, 29(9):1461–1475.

Alam, K. C. A., Kang, Y. T., Saha, B. B., Akisawa, A., and Kashiwagi, T. (2003). A novel approach to determine optimum switching frequency of a conventional adsorption chiller. *Energy*, 28(10):1021–1037.

Alam, K. C. A., Khan, M., Uyun, A. S., Hamamoto, Y., Akisawa, A., and Kashiwagi, T. (2007). Experimental study of a low temperature heat driven re-heat two-stage adsorption chiller. *Applied Thermal Engineering*, 27(10):1686–1692.

Aristov, Y. I. (2013a). Challenging offers of material science for adsorption heat transformation: A review. *Applied Thermal Engineering*, 50(2):1610–1618.

Aristov, Y. I. (2013b). Experimental and numerical study of adsorptive chiller dynamics: Loose grains configuration. *Applied Thermal Engineering*, 61(2):841–847.

Aristov, Y. I. (2014a). Adsorption Dynamics in Adsorptive Heat Transformers: Review of New Trends. *Heat Transfer Engineering*, 35(11-12):1014–1027.

Aristov, Y. I. (2014b). Concept of adsorbent optimal for adsorptive cooling/heating. *Applied Thermal Engineering*, 72(2):166–175.

Aristov, Y. I., Dawoud, B., Glaznev, I. S., and Elyas, A. (2008). A new methodology of studying the dynamics of water sorption/desorption under real operating conditions of adsorption heat pumps: Experiment. *International Journal of Heat and Mass Transfer*, 51(19-20):4966–4972.

Aristov, Y. I., Glaznev, I. S., and Girnik, I. S. (2012a). Optimization of adsorption dynamics in adsorptive chillers: Loose grains configuration. *Energy*, 46(1):484–492.

Aristov, Y. I., Sapienza, A., Ovoshchnikov, D. S., Freni, A., and Restuccia, G. (2012b). Reallocation of adsorption and desorption times for optimisation of cooling cycles. *International Journal of Refrigeration*, 35(3):525–531.

Bathen, D. (2001). *Adsorptionstechnik (in German)*. VDI-Buch. Springer, Berlin, Heidelberg.

Bau, U., Braatz, A.-L., Lanzerath, F., Herty, M., and Bardow, A. (2015a). Control of adsorption chillers by a gradient descent method for optimal cycle time allocation. *International Journal of Refrigeration*, 56:52–64.

Bau, U., Braatz, A.-L., Lanzerath, F., Herty, M., and Bardow, A. (2015b). Optimal Dynamic Operation of Adsorption-based Energy Systems Driven by Fluctuating Renewable Energy. In Gernaey, K. V., Huusom, J. K., and Gani, R., editors, *12th International Symposium on Process Systems Engineering and 25th European Symposium on Computer Aided Process Engineering*, volume 37 of *Computer-aided chemical engineering*, pages 2339–2344. Elsevier, Amsterdam.

Bau, U., Gibelhaus, A., Lanzerath, F., and Bardow, A. (2017a). Optimal operation of two-bed adsorption chillers by combining heat recovery and reallocation of adsorption/desorption times. In *Proceedings of International Sorption Heat Pump Conference 2017*.

Bau, U., Hoseinpoori, P., Graf, S., Schreiber, H., Lanzerath, F., and Bardow, A. (2016a). Rigorous assessment of adsorber-bed designs using dynamic optimization. In Eames, I. W. and Tierney, M. J., editors, *Proceedings of Heat Powered Cycles 2016*.

Bau, U., Hoseinpoori, P., Graf, S., Schreiber, H., Lanzerath, F., Kirches, C., and Bardow, A. (2017b). Dynamic optimisation of adsorber-bed designs ensuring optimal control. *Applied Thermal Engineering*, 125:1565–1576.

Bau, U., Lanzerath, F., and Bardow, A. (2017c). A simple heat-flow-based control strategy for one-bed adsorption chillers leading to Pareto-optimal performance. In *Proceedings of International Sorption Heat Pump Conference 2017*.

Bau, U., Lanzerath, F., Gräber, M., Schreiber, H., Thielen, N., and Bardow, A. (2014). Adsorption energy systems library - Modeling adsorption based chillers, heat pumps, thermal storages and desiccant systems. In Tummescheit, H. and Årzén, K.-E., editors, *Proceedings of the 10th International Modelica Conference*, Linköping electronic conference proceedings, pages 875–883, Linköping. Modelica Association.

Bau, U., Neitzke, D., Lanzerath, F., and Bardow, A. (2015c). Multi-objective optimization of dynamic systems combining genetic algorithms and Modelica: Application to adsorption air-conditioning systems. In Fritzson, P. and Elmqvist, H., editors, *Proceedings of the 11th International Modelica Conference*, Linköping electronic conference proceedings, pages 777–784, Linköping. Modelica Association.

Bau, U., Schreiber, H., Lanzerath, F., and Bardow, A. (2015d). Adsorption-based air-conditioning for battery-driven electric busses. In *Proceedings of the 24th IIR International Congress of Refrigeration*, Paris. International Institute of Refrigeration.

Bau, U., Schreiber, H., Lanzerath, F., and Bardow, A. (2016b). Hybride Klimatisierung für Elektrofahrzeuge: Kombination von Wärmepumpe und offenem Sorptionssystem (in German). In Deußen, N. and Steinberg, P., editors, *Wärmemanagement des Kraftfahrzeugs*, Haus-der-Technik Fachbuch.

Bau, U., Schreiber, H., Lanzerath, F., and Bardow, A. (2016c). Hybride Klimatisierung für Elektrofahrzeuge mit Wärmepumpe und offenem Sorptionssystem (in German). In Deutscher Kälte- und Klimatechnischer Verein (DKV) e.V., editor, *DKV-Tagung 2016*.

Bejan, A. (2013). *Convection heat transfer*. Wiley, Hoboken, N.J., 4th ed. edition.

Ben Amar, N., Sun, L.-M., and Meunier, F. E. (1996). Numerical analysis of adsorptive temperature wave regenerative heat pump. *Applications of Adsorption in Energy Transfer and Storage Symposium*, 16(5):405–418.

Betts, J. T. (2001). *Practical Methods for Optimal Control Using Nonlinear Programming*. SIAM, Philadelphia.

Binkert, J., Lauer, J., Diaconu, A., Ruß, W., Schreiber, H., and Bardow, A. (2013). Entwicklung einer Verfahrenskombination aus Zeolithwärmepumpe, Vakuumeindampfsystem und Blockheizkraftwerk zur energieeffizienten Wärmeversorgung von Brauereien (in German).

Bock, H. G. and Plitt, K. J. (1984). A Multiple Shooting algorithm for direct solution of optimal control problems. In *Proceedings of the 9th IFAC World Congress*, pages 242–247, Budapest. Pergamon Press.

Bryson, A. E. and Ho, Y.-C. (1975). *Applied Optimal Control*. Wiley, New York.

Büttner, T. and Mittelbach, W. (2013). Method for controlling the power of a sorption refrigeration system and device therefor (Patent US 12/735,159).

Cacciola, G. and Restuccia, G. (1995). Reversible adsorption heat pump: a thermodynamic model. *International Journal of Refrigeration*, 18(2):100–106.

Chan, K. C., Tso, C. Y., and Chao, C. Y. H. (2014). Study of heat and mass recovery process on the performance of adsorption cooling systems. In *HEFAT2014: 10th International Conference on Heat Transfer, Fluid Mechanics and Thermodynamics*, pages 557–563, Orlando.

Chan, K. C., Tso, C. Y., Chao, C. Y. H., and Wu, C. L. (2015). Experiment verified simulation study of the operating sequences on the performance of adsorption cooling system. *Building Simulation*, 8(3):255–269.

Chang, K.-S., Chen, M.-T., and Chung, T.-W. (2005). Effects of the thickness and particle size of silica gel on the heat and mass transfer performance of a silica gel-coated bed for air-conditioning adsorption systems. *Applied Thermal Engineering*, 25(14-15):2330–2340.

Chang, W.-S., Wang, C.-C., and Shieh, C.-C. (2007). Experimental study of a solid adsorption cooling system using flat-tube heat exchangers as adsorption bed. *Applied Thermal Engineering*, 27(13):2195–2199.

Chang, W.-S., Wang, C.-C., and Shieh, C.-C. (2009). Design and performance of a solar-powered heating and cooling system using silica gel/water adsorption chiller. *Applied Thermal Engineering*, 29(10):2100–2105.

Cho, S.-H. and Kim, J.-N. (1992). Modeling of a silica gel/water adsorption-cooling system. *Energy*, 17(9):829–839.

Choudhury, B., Saha, B. B., Chatterjee, P. K., and Sarkar, J. P. (2013). An overview of developments in adsorption refrigeration systems towards a sustainable way of cooling. *Applied Energy*, 104:554–567.

Chua, H. T., Ng, K. C., Malek, A., Kashiwagi, T., Akisawa, A., and Saha, B. B. (1999). Modeling the performance of two-bed, sillica gel-water adsorption chillers. *International Journal of Refrigeration*, 22(3):194–204.

Chua, H. T., Ng, K. C., Malek, A., Kashiwagi, T., Akisawa, A., and Saha, B. B. (2001). Multi-bed regenerative adsorption chiller — improving the utilization of waste heat and reducing the chilled water outlet temperature fluctuation. *International Journal of Refrigeration*, 24(2):124–136.

Chua, H. T., Ng, K. C., Wang, W., Yap, C., and Wang, X. L. (2004). Transient modeling of a two-bed silica gel–water adsorption chiller. *International Journal of Heat and Mass Transfer*, 47(4):659–669.

Critoph, R. E., Metcalf, S. J., and Rivero-Pacho, A. M. (2016). Simulation of adsorption generators for optimal preformance. In Eames, I. W. and Tierney, M. J., editors, *Proceedings of Heat Powered Cycles 2016*.

Daou, K., Wang, R. Z., Xia, Z. Z., and Yang, G. Z. (2007). Experimental comparison of the sorption and refrigerating performances of a CaCl2 impregnated composite adsorbent and those of the host silica gel. *International Journal of Refrigeration*, 30(1):68–75.

de Oliveira, R. G. and Generoso, D. J. (2016). Influence of the operational conditions on the performance of a chemisorption chiller driven by hot water between 65 °C and 80 °C. *Applied Energy*, 162:257–265.

Deng, J., Wang, R. Z., and Han, G. (2011). A review of thermally activated cooling technologies for combined cooling, heating and power systems. *Progress in Energy and Combustion Science*, 37(2):172–203.

Diehl, M. (2011). *Numerical Optimal Control*. Script, ESAT-SCD, K.U. Leuven,, Leuven.

Diehl, M., Leineweber, D. B., and Schäfer, A. A. S. (2001). MUSCOD-II Users' Manual: IWR-Pre-print.

Douss, N., Meunier, F. E., and Sun, L.-M. (1988). Predictive model and experimental results for a two-adsorber solid adsorption heat pump. *Industrial & Engineering Chemistry Research*, 27(2):310–316.

Dubinin, M. M. (1960). Theory of the physical adsorption of gases and vapors and adsorption properties of adsorbents of various natures and porous structures. *Russian Chemical Bulletin*, 9:1072–1078.

Dubinin, M. M. (1967). Adsorption in micropores. *Journal of Colloid and Interface Science*, 23(4):487–499.

Duong, D. D. (1998). *Adsorption analysis: Equilibria and kinetics*, volume 2 of *Series on chemical engineering*. Imperial College Press, London.

Ehrgott, M. (2010). *Multicriteria optimization*. Springer, Berlin and Heidelberg, 2nd edition.

El-Sharkawy, I. I., AbdelMeguid, H., and Saha, B. B. (2013). Towards an optimal performance of adsorption chillers: Reallocation of adsorption/desorption cycle times. *International Journal of Heat and Mass Transfer*, 63:171–182.

Erdogan, M., Graf, S., Bau, U., Lanzerath, F., and Bardow, A. (2017). Simple two-step assessment of novel adsorbents for drying: The trade-off between adsorber size and drying time. *Applied Thermal Engineering*, 125:1075–1082.

Frazzica, A., Füldner, G., Sapienza, A., Freni, A., and Schnabel, L. (2014). Experimental and theoretical analysis of the kinetic performance of an adsorbent coating composition for use in adsorption chillers and heat pumps. *Applied Thermal Engineering*, 73(1):1022–1031.

Frazzica, A., Palomba, V., Dawoud, B., Gullì, G., Brancato, V., Sapienza, A., Vasta, S., Freni, A., Costa, F., and Restuccia, G. (2016). Design, realization and testing of an adsorption refrigerator based on activated carbon/ethanol working pair. *Applied Energy*, 174:15–24.

Freni, A. (2015). *Characterization of zeolite-based coatings for adsorption heat pumps*. SpringerBriefs in applied sciences and technology. Springer, Cham.

Freni, A., Bonaccorsi, L., Calabrese, L., Caprì, A., Frazzica, A., and Sapienza, A. (2015). SAPO-34 coated adsorbent heat exchanger for adsorption chillers. *Applied Thermal Engineering*, 82:1–7.

Freni, A., Maggio, G., Cipitì, F., and Aristov, Y. I. (2012a). Simulation of water sorption dynamics in adsorption chillers: One, two and four layers of loose silica grains. *Applied Thermal Engineering*, 44:69–77.

Freni, A., Sapienza, A., Glaznev, I. S., Aristov, Y. I., and Restuccia, G. (2012b). Experimental testing of a lab-scale adsorption chiller using a novel selective water sorbent "silica modified by calcium nitrate". *International Journal of Refrigeration*, 35(3):518–524.

Fritzson, P. (2015). *Principles of object oriented modeling and simulation with Modelica 3.3: A cyber-physical approach*. John Wiley & Sons Inc, Hoboken, New Jersey, 2nd edition.

Frolov, T. and Mishin, Y. (2015). Phases, phase equilibria, and phase rules in low-dimensional systems. *The Journal of Chemical Physics*, 143(4):044706.

Gerdts, M. (2012). *Optimal Control of ODEs and DAEs*. De Gruyter textbook. De Gruyter, Berlin.

Gibelhaus, A., Fidorra, N., Lanzerath, F., Bau, U., Schnabel, L., Köhler, J., and Bardow, A. (2017). Hybrid refrigeration with CO2 vapor compression cycle and adsorption chiller: An efficient combination of natural working fluids. In *Proceedings of International Sorption Heat Pump Conference 2017*.

Giraud, F., Rullière, R., Toublanc, C., Clausse, M., and Bonjour, J. (2015). Experimental evidence of a new regime for boiling of water at subatmospheric pressure. *Experimental Thermal and Fluid Science*, 60:45–53.

Glueckauf, E. (1955). Theory of Chromatography: Part 10: Formulae for Diffusion into Spheres and their Application to Chromatography. *Transactions of the Faraday Society*, 51(11):1540–1551.

Goldsworthy, M. J. (2014). Measurements of water vapour sorption isotherms for RD silica gel, AQSOA-Z01, AQSOA-Z02, AQSOA-Z05 and CECA zeolite 3A. *Microporous and Mesoporous Materials*, 196:59–67.

Gong, L. X., Wang, R. Z., Xia, Z. Z., and Chen, C. J. (2011). Design and performance prediction of a new generation adsorption chiller using composite adsorbent. *Energy Conversion and Management*, 52(6):2345–2350.

Gong, L. X., Wang, R. Z., Xia, Z. Z., and Lu, Z. S. (2012). Experimental study on an adsorption chiller employing lithium chloride in silica gel and methanol. *International Journal of Refrigeration*, 35(7):1950–1957.

Gräber, M. (2013). *Energieoptimale Regelung von Kälteprozessen (in German)*. PhD thesis, Technische Universität Carolo-Wilhelmina zu Braunschweig, Braunschweig.

Gräber, M., Kirches, C., Bock, H. G., Schlöder, J. P., Tegethoff, W., and Köhler, J. (2011). Determining the optimum cyclic operation of adsorption chillers by a direct method for periodic optimal control. *International Journal of Refrigeration*, 34(4):902–913.

Gräber, M., Kirches, C., Scharff, D., and Tegethoff, W. (2012). Using Functional Mock-up Units for Nonlinear Model Predictive Control. In Otter, M. and Zimmer, D., editors, *9th International Modelica Conference*, Linköping electronic conference proceedings, pages 781–790, Linköping. Linköping University Electronic Press.

Gräber, M., Kosowski, K., Richter, C., and Tegethoff, W. (2010). Modelling of heat pumps with an object-oriented model library for thermodynamic systems. *Mathematical and Computer Modelling of Dynamical Systems*, 16(3):195–209.

Graf, S., Bau, U., Erdogan, M., and Bardow, A. (2017a). The optimal adsorbent for adsorption chiller cycles. In *Proceedings of International Sorption Heat Pump Conference 2017*.

Graf, S., Becker, D., Ackermann, J., Lanzerath, F., and Bardow, A. (2015). Heat and mass transfer mechanisms in adsorption heat pumps: Experiment and dynamic modeling. In *IX Minsk International Seminar "Heat Pipes, Heat Pumps, Refrigerators, Power Sources"*, volume 1, pages 264–274.

Graf, S., Lanzerath, F., and Bardow, A. (2017b). The IR-Large-Temperature-Jump method: Determining heat and mass transfer coefficients for adsorptive heat transformers. *Applied Thermal Engineering*, 126:630–642.

Graf, S., Lanzerath, F., Sapienza, A., Frazzica, A., Freni, A., and Bardow, A. (2016). Prediction of SCP and COP for adsorption heat pumps and chillers by combining the large-temperature-jump method and dynamic modeling. *Applied Thermal Engineering*, 98:900–909.

Grisel, R. J., Smeding, S. F., and Boer, R. d. (2010). Waste heat driven silica gel/water adsorption cooling in trigeneration. *Applied Thermal Engineering*, 30(8-9):1039–1046.

Haimes, Y., Lasdon, L., and Wismer, D. (1971). On a bicriterion formulation of the problems of integrated system identification and system optimization. *IEEE Transactions on Systems, Man, and Cybernetics*, 1(3):296–297.

Hamamoto, Y., Alam, K. C. A., Akisawa, A., and Kashiwagi, T. (2005). Performance evaluation of a two-stage adsorption refrigeration cycle with different mass ratio. *International Journal of Refrigeration*, 28(3):344–352.

Holland, J. H. (1975). *Adaptation in natural and artificial systems: An introductory analysis with applications to biology, control, and artificial intelligence.* University of Michigan Press, Ann Arbor.

Hougen, O. A. and Marshall, W. R. (1947). Adsorption from a fluid stream flowing through a stationary granular bed. *Chemical Engineering Progress*, (43):197–208.

Hsu, C. T., Cheng, P., and Wong, K. W. (1994). Modified Zehner-Schlunder models for stagnant thermal conductivity of porous media. *International Journal of Heat and Mass Transfer*, 37(17):2751–2759.

Huangfu, Y., Wu, J. Y., Wang, R. Z., and Xia, Z. Z. (2007). Experimental investigation of adsorption chiller for Micro-scale BCHP system application. *Energy and Buildings*, 39(2):120–127.

International Energy Agency (2014). Heating without Global Warming: Market Developments and Policy Considerations for Renewable Heat.

Joint Committee for Guides in Metrology (2008). Evaluation of measurement data - Guide to the expression of uncertainty in measurement.

Jones, D., Mirrazavi, S., and Tamiz, M. (2002). Multi-objective meta-heuristics: An overview of the current state-of-the-art. *European Journal of Operational Research*, 137(1):1–9.

Joos, A., Dietl, K., and Schmitz, G. (2009). Thermal Separation: An Approach for a Modelica Library for Absorption, Adsorption and Rectification. In Casella, F., editor, *Proceedings of the 7th International Modelica Conference*, Linköping electronic conference proceedings, pages 804–813, Linköping. Linköping University Electronic Press.

Kennedy, J. and Russell, E. (1995). Particle Swarm Optimization. In IEEE, editor, *Proceedings / 1995 IEEE International Conference on Neural Networks*, pages 1942–1948, Piscataway, NJ. IEEE Service Center.

Körkel, S., Qu, H., Rücker, G., and Sager, S. (2005). Derivative Based vs. Derivative Free Optimization Methods for Nonlinear Optimum Experimental Design. In Zhang, W., Tong, W., Chen, Z., and Glowinski, R., editors, *Current Trends in High Performance Computing and Its Applications*, pages 339–344. Springer-Verlag, Berlin/Heidelberg.

Krstic, M. and Smyshlyaev, A. (2008). *Boundary control of PDEs*, volume 16 of *Advances in Design and Control*. Society for Industrial and Applied Mathematics (SIAM), Philadelphia, PA.

Kubota, M., Ueda, T., Fujisawa, R., Kobayashi, J., Watanabe, F., Kobayashi, N., and Hasatani, M. (2008). Cooling output performance of a prototype adsorption heat pump with fin-type silica gel tube module. *Applied Thermal Engineering*, 28(2-3):87–93.

Lanzerath, F. (2014). *Modellgestützte Entwicklung von Adsorptionswärmepumpen (in German)*, volume 3 of *Aachener Beiträge zur Technischen Thermodynamik*. Mainz, Aachen.

Lanzerath, F., Bau, U., Seiler, J., and Bardow, A. (2015). Optimal design of adsorption chillers based on a validated dynamic object-oriented model. *Science and Technology for the Built Environment*, 21(3):248–257.

Lanzerath, F., Seiler, J., Bau, U., and Bardow, A. (2014). A modular experimental and simulation approach for the systematic development of adsorption heat pumps. In Radermacher, R., editor, *International Sorption Heat Pump Conference (ISHPC 2014)*, pages 117–126. Curran, Red Hook, NY.

Lanzerath, F., Seiler, J., Erdogan, M., Schreiber, H., Steinhilber, M., and Bardow, A. (2016). The impact of filling level resolved: Capillary-assisted evaporation of water for adsorption heat pumps. *Applied Thermal Engineering*, 102:513–519.

Leineweber, D. B., Bauer, I., Bock, H. G., and Schlöder, J. P. (2003). An efficient multiple shooting based reduced SQP strategy for large-scale dynamic process optimization (Parts I and II). *Computers & Chemical Engineering*, 27(2):157–174.

Lemmon, E. W., Huber, M. L., and McLinden, M. O. (2013). *NIST Standard Reference Database 23*.

Leong, K. C. and Liu, Y. (2004a). Numerical modeling of combined heat and mass transfer in the adsorbent bed of a zeolite/water cooling system. *Applied Thermal Engineering*, 24(16):2359–2374.

Leong, K. C. and Liu, Y. (2004b). Numerical study of a combined heat and mass recovery adsorption cooling cycle. *International Journal of Heat and Mass Transfer*, 47(22):4761–4770.

Li, S. L., Xia, Z. Z., Wu, J. Y., Li, J., Wang, R. Z., and Wang, L. W. (2010). Experimental study of a novel CaCl2/expanded graphite-NH3 adsorption refrigerator. *International Journal of Refrigeration*, 33(1):61–69.

Li, X. H., Hou, X. H., Zhang, X., and Yuan, Z. X. (2015). A review on development of adsorption cooling—Novel beds and advanced cycles. *Energy Conversion and Management*, 94:221–232.

Liu, Y. L., Wang, R. Z., and Xia, Z. Z. (2005a). Experimental performance of a silica gel–water adsorption chiller. *Applied Thermal Engineering*, 25(2-3):359–375.

Liu, Y. L., Wang, R. Z., and Xia, Z. Z. (2005b). Experimental study on a continuous adsorption water chiller with novel design. *International Journal of Refrigeration*, 28(2):218–230.

Magnetto, D. (2005). TOPMACS: Thermally Operated Mobile Air Conditioning Systems.

Metcalf, S. J., Critoph, R. E., and Tamainot-Telto, Z. (2012). Optimal cycle selection in carbon-ammonia adsorption cycles. *International Journal of Refrigeration*, 35(3):571–580.

Meunier, F. E. (1985). Second law analysis of a solid adsorption heat pump operating on a reversible cascade cycles: Application to the zeolite-water pair. *Heat Recovery Systems*, 5(2):133–141.

Meunier, F. E. (2013). Adsorption heat powered heat pumps. *Applied Thermal Engineering*, 61(2):830–836.

Meunier, F. E., Neveu, P., and Castaing-Lasvignottes, J. (1998). Equivalent Carnot cycles for sorption refrigeration. *International Journal of Refrigeration*, 21(6):472–489.

Miles, D. J. and Shelton, S. V. (1996). Design and testing of a solid-sorption heat-pump system. *Applied Thermal Engineering*, 16(5):389–394.

Miyazaki, T. and Akisawa, A. (2009). The influence of heat exchanger parameters on the optimum cycle time of adsorption chillers. *Applied Thermal Engineering*, 29(13):2708–2717.

Modelica Association Project "FMI" (2014). Functional Mockup Interface for Model Exchange and Co-Simulation, Version 2.0.

Nasruddin (2005). *Dynamic Modeling and Simulation of a Two-Bed Silicagel-Water Adsorption Chiller*. PhD Thesis, RWTH Aachen University, Aachen.

Ng, K. C., Wang, X., Lim, Y. S., Saha, B. B., Chakraborty, A., Koyama, S., Akisawa, A., and Kashiwagi, T. (2006). Experimental study on performance improvement of a four-bed adsorption chiller by using heat and mass recovery. *International Journal of Heat and Mass Transfer*, 49(19-20):3343–3348.

Nocedal, J. and Wright, S. J. (2006). *Numerical Optimization*. Springer, New York, 2nd edition.

Núñez, T. (2001). *Charakterisierung und Bewertung von Adsorbentien für Wärmetransformationsanwendungen*. PhD Thesis, Albert-Ludwigs-Universität, Freiburg.

Núñez, T., Mittelbach, W., and Henning, H.-M. (2007). Development of an adsorption chiller and heat pump for domestic heating and air-conditioning applications. *Applied Thermal Engineering*, 27(13):2205–2212.

Okunev, B. N. and Aristov, Y. I. (2014). Making adsorptive chillers faster by a proper choice of adsorption isobar shape: Comparison of optimal and real adsorbents. *Energy*, 76:400–405.

Ouchlyama, N. and Tanaka, T. (1984). Porosity Estimation for Random Packings of Spherical Particles. *Industrial and Engineering Chemistry Fundamentals*, 23:490–493.

Palomba, V., Dawoud, B., Sapienza, A., Vasta, S., and Frazzica, A. (2017). On the impact of different management strategies on the performance of a two-bed activated carbon/ethanol refrigerator: An experimental study. *Energy Conversion and Management*, 142:322–333.

Pan, Q. W., Wang, R. Z., Lu, Z. S., and Wang, L. W. (2014). Experimental investigation of an adsorption refrigeration prototype with the working pair of composite adsorbent-ammonia. *Applied Thermal Engineering*, 72(2):275–282.

Pesaran, A., Lee, H., Hwang, Y., Radermacher, R., and Chun, H.-H. (2016). Review article: Numerical simulation of adsorption heat pumps. *Energy*, 100:310–320.

Pesaran, A. A. and Mills, A. F. (1987a). Moisture transport in silica gel packed beds—II. Experimental study. *International Journal of Heat and Mass Transfer*, 30(6):1051–1060.

Pesaran, A. A. and Mills, A. F. (1987b). Moisture transport in silica gel packed beds—I.Theoretical study. *International Journal of Heat and Mass Transfer*, 30(6):1037–1049.

Polanyi, M. (1916). Adsorption of gases by a non-volatile adsorbent. *Verhandlungen der Deutschen Physikalischen Gesellschaft*, 18(2-3):55–80.

Pons, M. (1996). Second law analysis of adsorption cycles with thermal regeneration. *Journal of Energy Resources Technology*, 118:229–236.

Pons, M. and Feng, Y. (1997). Characteristic parameters of adsorptive refrigeration cycles with thermal regeneration. *Applied Thermal Engineering*, 17(3):289–298.

Pons, M. and Poyelle, F. (1999). Adsorptive machines with advanced cycles for heat pumping or cooling applications. *International Journal of Refrigeration*, 22(1):27–37.

Qian, S., Gluesenkamp, K., Hwang, Y., Radermacher, R., and Chun, H.-H. (2013). Cyclic steady state performance of adsorption chiller with low regeneration temperature zeolite. *Energy*, 60:517–526.

Rahman, A. F. M. M., Miyazaki, T., Ueda, Y., Saha, B. B., and Akisawa, A. (2013). Performance Comparison of Three-Bed Adsorption Cooling System With Optimal Cycle Time Setting. *Heat Transfer Engineering*, 34(11-12):938–947.

Rawlings, J. B. and Bakshi, B. R. (2006). Particle filtering and moving horizon estimation. *Computers & Chemical Engineering*, 30(10-12):1529–1541.

Rawlings, J. B. and Mayne, D. Q. (2009). *Model predictive control: Theory and design*. Nob Hill, Madison.

Rawlings, J. B. and Muske, K. R. (1993). The stability of constrained receding horizon control. *IEEE Trans. Automat. Control*, 38(10):1512–1516.

Restuccia, G., Freni, A., and Maggio, G. (2002). A zeolite-coated bed for air conditioning adsorption systems: parametric study of heat and mass transfer by dynamic simulation. *Applied Thermal Engineering*, 22(6):619–630.

Restuccia, G., Freni, A., Russo, F., and Vasta, S. (2005). Experimental investigation of a solid adsorption chiller based on a heat exchanger coated with hydrophobic zeolite. *Applied Thermal Engineering*, 25(10):1419–1428.

Restuccia, G., Freni, A., Vasta, S., and Aristov, Y. I. (2004). Selective water sorbent for solid sorption chiller: Experimental results and modelling. *International Journal of Refrigeration*, 27(3):284–293.

Rezk, A. R. and Al-Dadah, R. K. (2012). Physical and operating conditions effects on silica gel/water adsorption chiller performance. *Applied Energy*, 89(1):142–149.

Riffel, D. B., Wittstadt, U., Schmidt, F. P., Núñez, T., Belo, F. A., Leite, A. P., and Ziegler, F. (2010). Transient modeling of an adsorber using finned-tube heat exchanger. *International Journal of Heat and Mass Transfer*, 53(7-8):1473–1482.

Ruthven, D. M. (1984). *Principles of adsorption and adsorption processes.* Wiley, New York.

Saha, B. B., Boelman, E. C., and Kashiwagi, T. (1995). Computational analysis of an advanced adsorption-refrigeration cycle. *Energy*, 20(10):983–994.

Sapienza, A., Glaznev, I. S., Santamaria, S., Freni, A., and Aristov, Y. I. (2012). Adsorption chilling driven by low temperature heat: New adsorbent and cycle optimization. *Applied Thermal Engineering*, 32:141–146.

Sapienza, A., Gullì, G., Calabrese, L., Palomba, V., Frazzica, A., Brancato, V., La Rosa, D., Vasta, S., Freni, A., Bonaccorsi, L., and Cacciola, G. (2016). An innovative adsorptive chiller prototype based on 3 hybrid coated/granular adsorbers. *Applied Energy*, 179:929–938.

Sapienza, A., Santamaria, S., Frazzica, A., and Freni, A. (2011). Influence of the management strategy and operating conditions on the performance of an adsorption chiller. *Energy*, 36(9):5532–5538.

Sapienza, A., Santamaria, S., Frazzica, A., Freni, A., and Aristov, Y. I. (2014). Dynamic study of adsorbers by a new gravimetric version of the Large Temperature Jump method. *Applied Energy*, 113:1244–1251.

Schawe, D. (1999). *Theoretical and Experimental Investigations of an Adsorption Heat Pump with Heat Transfer between two Adsorbers.* PhD Thesis, Universität Stuttgart, Stuttgart.

Schicktanz, M. (2011). Influence of temperature fluctuations on the operational behaviour of adsorption chillers. In Lazzarin, R. M., Longo, G. A., and Noro, M., editors, *International Sorption Heat Pump Conference (ISHPC 11)*, pages 351–357, Paris. International Institute of Refrigeration.

Schicktanz, M. and Núñez, T. (2009). Modelling of an adsorption chiller for dynamic system simulation. *International Journal of Refrigeration*, 32(4):588–595.

Schmidt, E. F. (1967). Wärmeübergang und Druckverlust in Rohrschlangen (in German). *Chemie Ingenieur Technik - CIT*, 39(13):781–789.

Schreiber, H. (2017). *Experiments and Validated Models for Adsorption Thermal Energy Storage in Industrial and Residential Applications*. PhD Thesis, RWTH Aachen University, Aachen.

Schreiber, H., Bau, U., Lanzerath, F., and Bardow, A. (2016a). Heat loss in adsorption thermal energy storage: modeling and experimental validation. In Eames, I. W. and Tierney, M. J., editors, *Proceedings of Heat Powered Cycles 2016*.

Schreiber, H., Graf, S., Lanzerath, F., and Bardow, A. (2014). Adsorption heat storage for combined heat and power units in industrial batch processes. In Radermacher, R., editor, *International Sorption Heat Pump Conference (ISHPC 2014)*. Curran, Red Hook, NY.

Schreiber, H., Graf, S., Lanzerath, F., and Bardow, A. (2015). Adsorption thermal energy storage for cogeneration in industrial batch processes: Experiment, dynamic modeling and system analysis. *Applied Thermal Engineering*, 89:485–493.

Schreiber, H., Lanzerath, F., Reinert, C., Grüntgens, C., and Bardow, A. (2016b). Heat lost or stored: Experimental analysis of adsorption thermal energy storage. *Applied Thermal Engineering*, 106:981–991.

Sharafian, A. and Bahrami, M. (2014). Assessment of adsorber bed designs in waste-heat driven adsorption cooling systems for vehicle air conditioning and refrigeration. *Renewable and Sustainable Energy Reviews*, 30:440–451.

Sharafian, A. and Bahrami, M. (2015). Critical analysis of thermodynamic cycle modeling of adsorption cooling systems for light-duty vehicle air conditioning applications. *Renewable and Sustainable Energy Reviews*, 48:857–869.

Sharafian, A., Dan, P. C., Huttema, W., and Bahrami, M. (2016a). Performance analysis of a novel expansion valve and control valves designed for a waste heat-driven two-adsorber bed adsorption cooling system. *Applied Thermal Engineering*, 100:1119–1129.

Sharafian, A., Nemati Mehr, S. M., Huttema, W., and Bahrami, M. (2016b). Effects of different adsorber bed designs on in-situ water uptake rate measurements of AQSOA FAM-Z02 for vehicle air conditioning applications. *Applied Thermal Engineering*, 98:568–574.

Sieder, E. N. and Tate, G. E. (1936). Heat Transfer and Pressure Drop of Liquids in Tubes. *Industrial & Engineering Chemistry*, 28(12):1429–1435.

Solmuş, İ., Kaftanoğlu, B., Yamalı, C., and Baker, D. (2011). Experimental investigation of a natural zeolite–water adsorption cooling unit. *Applied Energy*, 88(11):4206–4213.

Tamainot-Telto, Z. and Critoph, R. E. (2003). Advanced solid sorption air conditioning modules using monolithic carbon–ammonia pair. *Applied Thermal Engineering*, 23(6):659–674.

Teng, W. S., Leong, K. C., and Chakraborty, A. (2016). Revisiting adsorption cooling cycle from mathematical modelling to system development. *Renewable and Sustainable Energy Reviews*, 63:315–332.

Tso, C. Y., Chan, K. C., and Chao, C. Y. H. (2014). Experimental investigation of a double-bed adsorption cooling system for application in green buildings. In *HEFAT2014: 10th International Conference on Heat Transfer, Fluid Mechanics and Thermodynamics*, pages 549–556, Orlando.

Tso, C. Y., Chan, K. C., Chao, C. Y. H., and Wu, C. L. (2015). Experimental performance analysis on an adsorption cooling system using zeolite 13X/CaCl2 adsorbent with various operation sequences. *International Journal of Heat and Mass Transfer*, 85:343–355.

Tso, C. Y., Chan, K. C., Wu, C. L., and Chao, C. Y. H. (2016). Experimental Investigation of a Zeolite 13X/CaCl2 – Water Adsorption Cooling System. In Heiselberg, P. K., editor, *CLIMA 2016 - proceedings of the 12th REHVA World Congress*, Aalborg. Aalborg University, Department of Civil Engineering.

Tummescheit, H. (2002). *Design and Implementation of Object-Oriented Model Libraries using Modelica*. PhD Thesis, Lund University, Lund.

United Nations (2015). General Assembly resolution 70/1: Transforming our world: the 2030 Agenda for Sustainable Development: A/RES/70/1.

United Nations / Framework Convention on Climate Change (2015). Adoption of the Paris Agreement.

Vasta, S., Freni, A., Sapienza, A., Costa, F., and Restuccia, G. (2012). Development and lab-test of a mobile adsorption air-conditioner. *International Journal of Refrigeration*, 35(3):701–708.

Vasta, S., Melograno, P. N., Palomba, V., Sapienza, A., Motta, M., and Freni, A. (2016). Validation of an assessment procedure for adsorption chillers installed in solar cooling plants. In Eames, I. W. and Tierney, M. J., editors, *Proceedings of Heat Powered Cycles 2016*.

Veselovskaya, J. V., Critoph, R. E., Thorpe, R. N., Metcalf, S. J., Tokarev, M. M., and Aristov, Y. I. (2010). Novel ammonia sorbents "porous matrix modified by active salt" for adsorptive heat transformation: 3. Testing of "BaCl2/vermiculite" composite in a lab-scale adsorption chiller. *Applied Thermal Engineering*, 30(10):1188–1192.

Voyiatzis, E., Palyvos, J. A., and Markatos, N.-C. (2008). Heat-exchanger design and switching-frequency effects on the performance of a continuous type solar adsorption chiller. *Applied Energy*, 85(12):1237–1250.

Wang, D., Zhang, J., Tian, X., Liu, D., and Sumathy, K. (2014a). Progress in silica gel–water adsorption refrigeration technology. *Renewable and Sustainable Energy Reviews*, 30(0):85–104.

Wang, D. C., Shi, Z. X., Yang, Q. R., Tian, X. L., Zhang, J. C., and Wu, J. Y. (2007). Experimental research on novel adsorption chiller driven by low grade heat source. *Energy Conversion and Management*, 48(8):2375–2381.

Wang, D. C., Wang, Y. J., Zhang, J. P., Tian, X. L., and Wu, J. Y. (2008). Experimental study of adsorption chiller driven by variable heat source. *Energy Conversion and Management*, 49(5):1063–1073.

Wang, D. C., Xia, Z. Z., Wu, J. Y., Wang, R. Z., Zhai, H., and Dou, W. D. (2005a). Study of a novel silica gel–water adsorption chiller. Part I. Design and performance prediction. *International Journal of Refrigeration*, 28(7):1073–1083.

Wang, J., Wang, L. W., Luo, W. L., and Wang, R. Z. (2013). Experimental study of a two-stage adsorption freezing machine driven by low temperature heat source. *International Journal of Refrigeration*, 36(3):1029–1036.

Wang, L. W., Wang, R. Z., and de Oliveira, R. G. (2009). A review on adsorption working pairs for refrigeration. *Renewable and Sustainable Energy Reviews*, 13(3):518–534.

Wang, R. Z., Wang, L., and Wu, J. (2014b). *Adsorption refrigeration technology: Theory and application*. John Wiley & Sons, Singapore.

Wang, X. and Chua, H. T. (2007). A comparative evaluation of two different heat-recovery schemes as applied to a two-bed adsorption chiller. *International Journal of Heat and Mass Transfer*, 50(3-4):433–443.

Wang, X., Chua, H. T., and Ng, K. C. (2005b). Experimental investigation of silica gel–water adsorption chillers with and without a passive heat recovery scheme. *International Journal of Refrigeration*, 28(5):756–765.

Wittstadt, U., Jahnke, A., Schnabel, L., Sosnowski, M., Schmidt, F. P., and Ziegler, F. (2008). Test facility for small-scale adsorbers. In *International Sorption Heat Pump Conference 2008*.

Xia, Z. Z., Wang, D., and Zhang, J. (2008). Experimental study on improved two-bed silica gel–water adsorption chiller. *Energy Conversion and Management*, 49(6):1469–1479.

Xia, Z. Z., Yang, G. Z., and Wang, R. Z. (2009). Capillary-assisted flow and evaporation inside circumferential rectangular micro groove. *International Journal of Heat and Mass Transfer*, 52(3-4):952–961.

Yang, G. Z., Xia, Z. Z., Wang, R. Z., Keletigui, D., Wang, D. C., Dong, Z. H., and Yang, X. (2006). Research on a compact adsorption room air conditioner. *Energy Conversion and Management*, 47(15-16):2167–2177.

Yong, L. and Sumathy, K. (2002). Review of mathematical investigation on the closed adsorption heat pump and cooling systems. *Renewable and Sustainable Energy Reviews*, 6(4):305–338.

Zajaczkowski, B. (2016). Optimizing performance of a three-bed adsorption chiller using new cycle time allocation and mass recovery. *Applied Thermal Engineering*, 100:744–752.

Zhang, L. Z. and Wang, L. (1999). Effects of coupled heat and mass transfers in adsorbent on the performance of a waste heat adsorption cooling unit. *Applied Thermal Engineering*, 19(2):195–215.

Aachener Beiträge zur Technischen Thermodynamik

ABTT 1
Philip Voll
Automated Optimization-Based Synthesis of Distributed Energy Supply Systems
1. Auflage 2014
ISBN 978-3-86130-474-6

ABTT 2
Johannes Jung
Comparative Life Cycle Assessment of Industrial Multi-Product Processes
1. Auflage 2014
ISBN 978-3-86130-471-5

ABTT 3
Franz Lanzerath
Modellgestützte Entwicklung von Adsorptionswärmepumpen
1. Auflage 2014
ISBN 978-3-86130-472-2

ABTT 4
Thorsten Brands
Einfluss der Gemischzusammensetzung auf die Verbrennung im Diesel- und GCAI-Motor
1. Auflage 2014
ISBN 978-3-95886-006-3

ABTT 5
Dominique Dechambre
Efficient Measurement of Liquid-Liquid Equilibria using Automation and Optimal Experimental Design
1. Auflage 2016
ISBN 978-395886-077-3

ABTT 6
Niklas von der Aßen
From Life-Cycle Assesement towards life-Cycle Design of Carbon Dioxide Capture and Utilization
1. Auflage 2016
ISBN 978-3-95886-080-3

ABTT 7
Matthias Lampe
Integrated Process and Organic Rankine Cycle Working Fluid Design in the Continuous-Molecular Targeting Framework
1. Auflage 2016
ISBN 978-3-95886-086-5

ABTT 8
Thomas Hülser
Optische Untersuchung der Zündvorgänge und deren Auswirkung auf die Verbrennung in PKW-Motoren
1. Auflage 2016
ISBN 978-3-95886-090-2

Aachener Beiträge zur Technischen Thermodynamik

ABTT 9
Malte Döntgen
Reaction Models from Reactive Molecular Dynamics and High-Level Kinetics Predictions
1. Auflage 2016
ISBN 978-3-95886-156-5

ABTT 10
Heike Schreiber
Experiments and Validated Models for Adsorption Thermal Energy Storage in Industrial and Residential Application
1. Auflage 2017
ISBN 978-3-95886-178-7

ABTT 11
André Dirk Sternberg
System-Wide Perspective for Life Cycle Assesment of CO_2-based C1-Chemicals
1. Auflage 2017
ISBN 978-3-95886-193-0

ABTT 12
Uwe Bau
From Dynamic Simulation to Optimal Design and Control of Adsorption Energy Systems
1. Auflage 2018
ISBN 978-3-95886-216-6